I0648366

DOES GOD PLAY DICE?

For the author's comment on the title, see p. 122.

DOES GOD PLAY DICE?

A Look at the Story of the Universe

JOHN HOUGHTON
Foreword by Sir Robert Boyd

CANTILEVER BOOKS
Zondervan Publishing House
Grand Rapids, Michigan

CANTILEVER BOOKS is an imprint of Zondervan Publishing House, 1415 Lake Drive, S.E., Grand Rapids, Michigan 49506.

Library of Congress Cataloging in Publication Data

Houghton, John (John T.)
    Does God play dice? : a look at the story of the universe / John Houghton.
        p.   cm.
    Reprint. Originally published: Leicester, England : Inter-Varsity Press. c1988.
    Includes index.
    ISBN 0-310-51571-8
    1. Religion and science—1946-   I. Title.
    BL240.2.H69   1989
    261.5'5—dc20   89-9774
                                                                        CIP

Set in Garamond
Typeset in Great Britain by
Parker Typesetting Service, Leicester.
North American edition printed and bound by
Malloy Lithographing, Ann Arbor, Michigan.

Printed in the United States of America

89   90   91   92   93 / **ML** / 10   9   8   7   6   5   4   3   2   1

Contents

FOREWORD

by Sir Robert Boyd

When Laplace presented Napoleon with his abstruse 4,000-page thesis on celestial mechanics (or was it his later popular version, *Exposition du système du monde?*), he was asked what place was left for God in his account. Laplace is said to have replied, 'Sire, I have no need of that hypothesis.'

Why is it that many leading scientists, including John Houghton, claim that God is no mere hypothesis but a prime reality in their lives? It is certainly not that they would find themselves in disagreement with Laplace. Few would seek, as Newton did, to explain the inadequacies in their theories by acts of God. Ever since Laplace, it has been found increasingly that the fruitful response to gaps is more research. God has long since ceased to be a mere stop-gap.

Kant, whose 'theory of the heavens' had in some ways anticipated Laplace by half a century, had rejected the worship of God seen as a magician and had recognized the far more evocative concept of God as the giver of the Whole, physical laws and all. Yet here, two hundred years later, we in the West live in a society largely ignorant of both the aims and the limitations of science. On the one hand we are subjected by the media to a caricature of science which uses its authority to pronounce where its writ does not run. On the other hand, Christians, seduced by this propaganda, sadly all too often present us with a picture of a God so small that he is no more than a mere denizen of an autonomous universe. So far is he removed from the God of Christian theism that, like Newton's deistic God, his main function is the magical adjustment of space and time for which he no longer bears any other responsibility.

7

The biblical concept of God points to a far more organic relationship of the Creator to the Whole. Millennia ago a contributor to the book of Proverbs put it like this: 'The lot is cast into the lap, but its every decision is from the Lord' (16:33). To work out this relationship is not easy, any more than the pursuit of science is easy. Together, these difficulties account, in part at any rate, for the sloppy superstition of the poorly educated and the soft options selected by too many undergraduates.

Dr Houghton has sought, over many years, to integrate his science and his faith into a day-by-day practice. And it has not been easy. Life has brought both deep sadness and great joy. He carries the huge responsibility of Director General of the Meteorological Office with the same combination of scientific integrity and social responsibility as characterizes the thinking of this book.

When professor and head of the department of atmospheric physics at Oxford University, responsible for pioneering and trend-setting experiments in spacecraft on the atmospheres of planets, Dr Houghton still found time to run an active discussion group among his colleagues on the kinds of issue presented here.

When he was Director of the Science Research Council's Appleton Laboratory, Dr Houghton joined me in leading a delegation to discuss spacecraft collaboration with India. We arrived in Bombay early on a Sunday. It was characteristic of him to say, 'Robert, we don't have our first meeting until tomorrow. Let's spend some time in worship together.' At present I am privileged to represent the Royal Society on the country's Earth Observation Programme Board, of the British National Space Centre, over which he presides. It is a joy to see the same sense of stewardship of God's world colouring his chairmanship.

A foreword is no place in which to give a preview of a book. You have the book now in your hands. I commend it to you unhesitatingly, and I hope that many more, with you, will read it carefully and put your confidence in him of whom Paul wrote, 'He is the image of the invisible God . . . and in him all things hold together' (*Colossians 1:15–17*).

Sir Robert Boyd FRS
Emeritus Professor of Physics in the University of London
and formerly Director of the Mullard Space Science Laboratory
of University College

PREFACE

In order that we be whole human beings, I believe that different parts of our lives should relate together in an integrated way. The exploration of this book brings together, so far as I am able, two important strands of my life – namely my experience and career as a physicist and my experience as a Christian believer.

I have discussed the book's subject matter over the years with friends, colleagues and students; it is largely through such discussions that my thoughts and ideas have developed. In particular I am grateful to my first wife Margaret, who did not live to see the book published but without whose inspiration and encouragement it would never have been started. I am also grateful to the Rev. John Stott and the staff of the London Institute for Contemporary Christianity, who invited me to lecture on science and faith at a summer school in 1984, and to the students at that school. I also wish to thank those who have read and criticized the manuscript: in particular Dr Oliver Barclay, Sir Robert Boyd, Dr Paul Ewart, Dr Paul Fiddes, Dr Alan Gadd, Drs Janet and Paul Malcolm, the Rev. Jack Ramsbottom, Dr Barrie White and Professor John White.

Dr John Harries, Dr Bill Burton and Dr David Stickland of the Rutherford Appleton Laboratory assisted in providing some of the illustrations. Permission to reproduce illustrations has been given by Dr David Clark, Dr M. Seldner, the Lick Observatory, the European Southern Observatory, the Royal Greenwich Observatory and Messrs. Collins.

I am grateful to Mrs Val Jackman for typing the manuscript, to my wife Sheila for her continued help and suggestions and for providing the index, to Mr Peter Shirtcliffe for skilfully drawing the diagrams and to Mr Colin Duriez and the staff of Inter-Varsity Press for their help and cooperation in the production of the book.

John Houghton

Chapter One

THE SEARCH

Canst thou by searching find out God? (Job 11:7, Authorized Version)

In a small pea-sized lump of uranium of the kind which is used to make an atom bomb, about a thousand atoms emit radiation and decay every second. That is not very many compared with all the atoms in the lump which number about a hundred billion billion (10^{20});[1] only half the atoms will decay in a thousand million years. But which atoms will decay in the next second? A good question: the scientist cannot say; in fact, he will tell you quite categorically that there is no way of finding out.

The great physicist Albert Einstein was never very happy about the principles of chance and uncertainty which lie at the basis of our understanding of atoms and elementary particles and how they behave – so much so that he exclaimed, 'God does not play dice!' But why bring God into it? Does God know which atom is going to decay next? Need modern man with his scientific background give any thought to God? And what, if anything, has God to do with the universe, or with me? Can he be known? These are questions we all ask.

Thinking about God is perhaps the most challenging activity that can occupy our minds. We are looking for truth about God, the universe and nature. How can we set out on the search? Can science help, or must we rely solely on the application of faith? How different are the methods of science and faith in their search for truth? Are their methods related in any way?

Four different attitudes are currently held towards the relationship between these two methods of searching for truth. The first, perhaps the most common, is that the scientific method provides the only valid way of searching for truth. After all, the stuff of science, things we can see and handle, is much more in view than what appears to be the less tangible religious experience with which faith is concerned. An eminent scientist lecturing a few years ago in Oxford expressed just this view.[2] Religion, he pointed out, is not now taken seriously, and has by and large been displaced by science, the methods of which are proving so powerful and effective. Although he had no immediate concrete suggestions to offer, it is only therefore a matter of time, he argued, before science comes up with a replacement for religion. Such a view is common among scientists, and perhaps even more common among those unfamiliar with the scientific method.

Another attitude is the opposite of the first, namely that only faith is valid in the search for truth. Some, for instance, set the revelation of God found in the Bible over against the method of science, which is considered to be a man-made activity. Further, science, it is argued, is constantly changing; what scientists think today is different from what they thought fifty or a hundred years ago. Its changing character is set over against the unchanging revelation of God. Therefore, we are told, for whatever truth we are looking, be it ultimate scientific truth or be it truth about God, the revelation of God in Scripture is our only source.

A third common attitude is that science and faith are mutually exclusive. It was said (perhaps not completely fairly[3]) of Michael Faraday, the nineteenth-century physicist who was a regular local preacher, that when he left his laboratory he left his science behind, and that when he left the church, having preached, he left his faith behind. The two were kept apart. Much more recently, in 1972, the National Academy of Science of the USA, concerned about the California State Board of Education's proposal for parallel treatment to be given in school textbooks to the theory of evolution and to belief in special creation, resolved as follows: 'Whereas religion and science are therefore separate and mutually exclusive realms of human thought whose presentation in the same context leads to misunderstanding of the scientific theory and

12

religious belief ... we urge that the textbooks of science utilised in the public schools of the nation be limited to the exposition of scientific matters.'[4] We may well agree with their conclusion, but the reason they give for it illustrates all too well an attitude which is common for religious believers who are also scientists – that they carry out their research without reference to their faith and they exercise their faith without much relation to their scientific work.

But such an attitude is by no means held by all scientists who are also believers. Albert Einstein once said, 'Science without religion is lame, religion without science is blind,' illustrating a fourth possible attitude between science and faith. This is that the methods of science and of faith do overlap, and that we have experiences which relate to both, which can be explored by scientific means and also through the exercise of faith. This attitude is assumed in this book, and the exploration of the overlap is its purpose.

I need to make two particular points at the start. First, it is important to realize that the perspective from which faith is viewed is bound to be a somewhat personal one; hence a good deal of what I have to say inevitably has a personal flavour. Secondly, since I am a Christian, much of what I write is from the standpoint of the Christian faith.

A further point of explanation I need to make is that I have chosen the word 'faith' to identify what might otherwise be called the religious or the theological view. It may seem a surprising word to use, as in popular parlance it is often associated with credulity rather than with the critical approach of scientific discipline. I have, however, stuck with the word 'faith' and use it deliberately, because the words 'religion' or 'theology' have to do mainly with the *intellectual* study of religious material. It is necessary, in developing a balanced and realistic view of the interface between science and faith, not only to approach the evidence from an academic viewpoint but also to include the human response to that evidence from as many points of view as possible. *Faith* includes both the evidence and the response. It is also seen as the driving force for action.[5] This approach to the material of faith is paralleled in our approach to science, the study of which cannot be divorced from its applications and the way it works out in practice.

In brief outline, the exposition in the following chapters proceeds as follows. First, in order to present a broad scientific picture, we begin in chapter 2 with a look at the universe and the insight that science is providing into its origins and its evolution. We then go on to explore how God may be related to that scientific picture and to suggest scientific analogies or models which help our thinking about God and the universe. Later in the book, we go on further to explore relations between modes of thought in both science and faith, pointing out the importance of consistency in the overall perspectives of both science and faith. Finally, it is suggested that some degree of integration between the two perspectives can be achieved. As they are put alongside each other, a view of man's interaction with and responsibility for his environment can be developed – and, further, expressions of wonder at the scientific perspective can stimulate evocations of worship which in turn strengthen the perspective of faith.

---

[1] I shall conform to US usage which defines one billion as one thousand million (10^9).

[2] Sir George Porter in the Romanes Lecture, reported in the *Oxford Times*, 24 November 1978.

[3] C. A. Russell in *Cross-currents* (IVP, 1985) includes an analysis of Faraday's attitudes to science and faith.

[4] Quoted by A. R. Peacocke, *Creation and the World of Science* (Clarendon Press, 1978), p. 2.

[5] In Hebrews 11, notable actions based on faith are listed. The first verse of that chapter, 'Now faith is being sure of what we hope for and certain of what we do not see,' emphasizes the solid basis on which faith needs to rest.

Chapter Two

THE BIG BANG AND ALL THAT

Order is heaven's first law. (Alexander Pope)

We begin with a look at the universe. In the time taken to write this sentence, the furthest galaxies which can be seen with our telescopes have moved away from us another million miles. The universe is expanding; stars and galaxies are rushing apart. I say 'rushing apart' because they are moving at speeds which, by our everyday standards, are very high indeed. Yet, because of the vastness of space, these movements appear as very slow. The positions of the stars and constellations in the night sky are not noticeably different for us from what they were for the people who looked out on them at the beginning of recorded human history.

Just how vast is the universe as we now know it? Let us begin with our solar system, consisting of our sun with its nine planets revolving around it. The sun, some 1.5 million kilometres in diameter, is 150 million kilometres away from planet Earth and about 6 billion kilometres from Pluto, the furthest away of the nine planets.

Our sun is a modest member of a collection of stars known as a galaxy, of overall shape rather like a flat disc stretching across the sky in what is known as the Milky Way. Our sun resides in one of the galaxy's spiral arms about halfway from the centre to the edge. The whole galaxy, which is slowly rotating as it moves through space, is about a billion billion (10^{18}) kilometres across. The size of the galaxy is breathtaking enough; but even more breathtaking is the number of stars it

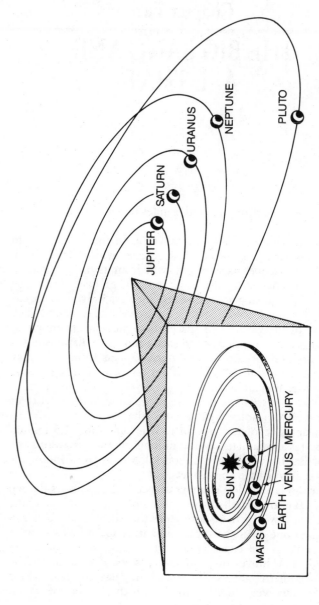

Fig. 2.1 *The planets in the solar system. The diameter of the orbit of Pluto is about 12 billion km.*

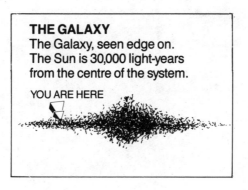

THE GALAXY
The Galaxy, seen edge on.
The Sun is 30,000 light-years
from the centre of the system.

YOU ARE HERE

Fig. 2.2 *The galaxy seen edge on. Our sun is 30,000 light years (3×10^{17} km) from the galaxy's centre.*

contains, about 100 billion (10^{11}) the nearest of which is some 40 million million (4×10^{13}) kilometres away.

Close by our galaxy are two smaller galaxies, the Magellanic Clouds, which appear as smudges in the southern-hemispheric night sky. Beyond them about thirty galaxies within 40 billion billion (4×10^{19}) kilometres of us form our local group of galaxies. Other groups and clusters of galaxies – upwards of a billion galaxies in all – are spread more or less uniformly through the universe; the furthest galaxies accessible to our telescopes are about 100,000 billion billion (10^{23}) kilometres away.

To try to appreciate these distances, imagine a scale model, of a size such that the sun and all the planets would fit into a typical house. The sun would be about the size of a pea; the earth and the other planets would be specks of dust. The nearest star would still be 100 kilometres away, the farthest edge of the galaxy well over a million kilometres and the edge of the universe a good fraction of a million million kilometres from us – still completely mind-boggling numbers.

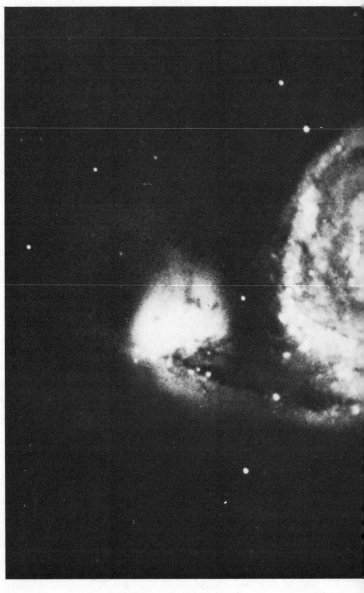

Fig. 2.3 *A spiral galaxy like our own, seen face on – Messier 51*

18

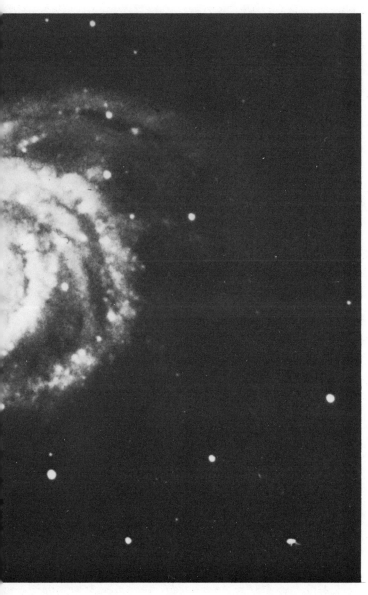

in Canes Venatici (from the Royal Greenwich Observatory).

Fig. 2.4 *Galaxy M31 in Andromeda, the nearest large galaxy
to our own. The two bright spots are smaller galaxies held by*

the gravitational field of M31 (from the Lick Observatory).

Because of these enormous numbers, astronomers measure distances in terms of the time it would take to travel along them moving at the speed of light (300,000 kilometres per second). At this speed, we would reach the sun in eight minutes; therefore the sun is said to be eight light minutes away. We would reach the nearest star in about four years (it is four light years away), the edge of our own galaxy in about 100,000 years, and the edge of the universe in about 10 billion (10^{10}) years.

Measuring distances in light years is a reminder that the light by which we see stars and galaxies has taken a long time to reach us. These objects are not, therefore, seen as they are now but as they were some time ago. We see the sun as it was eight minutes ago, and the nearest star as it was four years ago. When an explosion was noticed in the sky in the region of the Crab Nebula on 4 July 1054, the astronomers recording the explosion were seeing an event that occurred some 5,000 years earlier. A giant galaxy near to the edge of the universe viewed through one of our largest telescopes is seen as it was some 10 billion years ago. Since no information can travel faster than the speed of light, there is no means by which we can discover what has happened to these far off galaxies since then.

However, viewing the more distant parts of the universe as they were a long time ago has the advantage that it is possible to learn what these parts of the universe were like much earlier in the universe's history. In fact, by looking at different distances and therefore different periods back in time, some idea can be obtained of how the universe has evolved. Some of the evidence will be presented later in the chapter, but the conclusion which astronomers have reached as they have put together the historical jigsaw is that some 20 billion (2×10^{10}) or so years ago the matter and energy in the universe were all concentrated in a central mass at very high density, pressure and temperature, and that the expanding universe we see today started at that time with what has become known as the Big Bang.

Scientists from many parts of physics, from astronomy and from cosmology have all combined to piece together events since the Big Bang. Our present knowledge of physics does not allow us to go right back to the start, to time equal to zero. But we can come close to it.[1] A fraction of a second (perhaps about a hundredth of a second) after the start, matter and radiation at enormously high temperature (about 100 billion [10^{11}] degrees Celsius) and enormously high density (well over a billion times the density of water) made up the universe. A cricket ball of this material would weigh a million tons. The radiation was mostly what we call electromagnetic radiation which includes X-rays, ultraviolet light, visible light, infra-red radiation and radio waves. Radiation at these very high temperatures readily interchanges with matter in the form of a few elementary particles such as electrons. In the particle-and-energy 'soup', as it is colloquially called, no atoms or molecules could survive. Even the nuclei of simple atoms were broken down.

We have to wait for three or four minutes after the beginning when in the rapidly expanding universe the temperature is down to a mere billion (10^9) degrees (still much hotter than the centre of a hydrogen bomb) and the density perhaps ten times that of water. Under these conditions, the larger elementary particles, protons and neutrons can stick together to form simple nuclei; for instance, nuclei of helium, each with two protons and two neutrons – a particularly stable combination.

The expansion continues. Hydrogen and helium nuclei are present in abundance; so is radiation. It takes a million years or so for the universe to cool enough for electrons to attach themselves to the nuclei to form atoms. By this time, the universe is largely empty space filled with sparse, comparatively chilly clouds of hydrogen and helium gas at only a few atoms per cubic centimetre and a temperature of a few thousand degrees.

Imagine a region within these swirling clouds of higher density than the rest. The force of gravity will attract more matter into this more dense region. Higher-density blobs within this region where matter is concentrated will contract further. Over a period of millions of years, these high-density blobs will turn into stars, and groups of stars into galaxies.

As the hydrogen and helium gas concentrate into stars, and the pressure increases, so does the temperature (for the same reason as a bicycle pump becomes warm when air is compressed). Temperatures of millions of degrees Celsius are reached. Within these stellar furnaces, nuclear synthesis takes

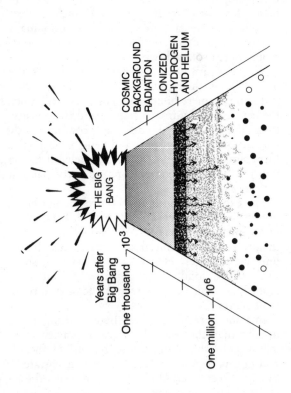

THE BIG BANG

COSMIC
BACKGROUND
RADIATION

IONIZED
HYDROGEN
AND HELIUM

Years after
Big Bang

One thousand —10^3

10^6

One million

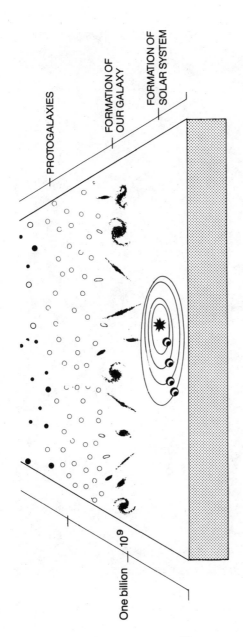

Fig. 2.5 *The evolution of the universe.*

PROTOGALAXIES

FORMATION OF OUR GALAXY

FORMATION OF SOLAR SYSTEM

One billion

10^9

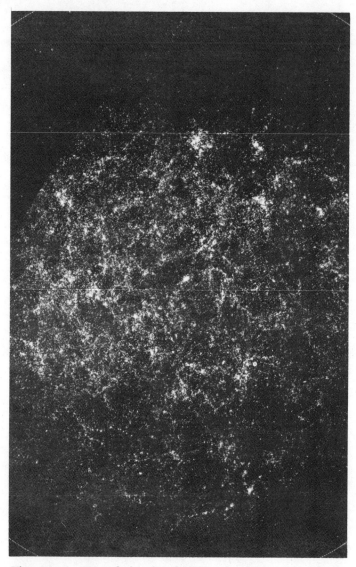

Fig. 2.6 *A map of the distribution of galaxies through the northern hemisphere sky (from Seldner* et al., Astronomical Journal *82 (1977), p. 313.*

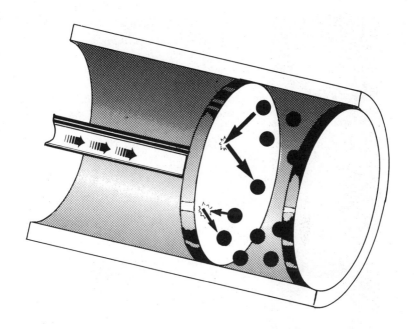

Fig. 2.7 *Gas molecules in a cylinder gain speed when they bounce off an advancing piston. This has the effect of heating the gas. In a similar way, the temperature inside a star increases as it contracts.*

place. Hydrogen nuclei fuse together to become helium, releasing a lot of energy in the process. As the hydrogen is used up, the star contracts and the temperature rises further. In these more extreme conditions, helium nuclei fuse to form carbon and oxygen. More complex nuclear reactions occur successively, forming heavier elements all the way up the periodic table to iron. The ambition of the alchemist to change elements from one form to another is in fact a continual cosmic process, occurring on an enormous scale within billions of stellar furnaces.

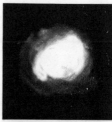

A sun-like star begins life as a cloud of interstellar material shrinking under the force of gravity. On becoming highly compressed, thermonuclear reactions start up which transform the star's hydrogen supply into helium – after 100 million years the star begins to shine.

The shining phase lasts for 10,000 million years.

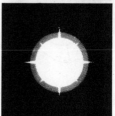

As the hydrogen is used up, the star swells up and shines more brightly as helium starts to burn.

Becoming unstable, the star sheds matter.

The star ends its life as a white dwarf.

Fig. 2.8 *The formation of stars*

(From David Clark, The Universe and Man, *Rutherford Appleton Laboratory, 1981)*

Even more extreme conditions are generated as some stars towards the end of their life blow themselves apart in events known as supernovae. We have already referred to such an explosion in the Crab constellation observed by astronomers in the year 1054; very recently, astronomers were excited when a new supernova was observed in the Larger Magellanic Cloud on 23 February 1987. (See pp. 48-49.) It is in these gigantic explosions, we believe, that heavy elements such as platinum, gold, uranium and a host of others are formed.

This exploded material, now containing quantities of all the ninety-two elements of the periodic table, mixes with hydrogen and helium gas of the interstellar medium in its turn to go again through the stellar evolutionary process. Second-generation stars are born containing debris from the disintegration of the first generation. We believe our sun to be such a second-generation star. Around our sun, planets have formed, probably as gas-and-dust clouds surrounding the young sun gradually fused together into a number of dense objects. So planet Earth was born some 4.5 billion years ago with its rich chemical composition and conditions suitable for the development of life.

Such is the story of the universe as currently understood, very briefly told. What is the evidence on which the story is based? Let me mention a few of the main building blocks.[2]

First, there is the rather obvious observation (whose significance was first pointed out by a German physician, Olbers, in 1826) that the sky at night is dark between the stars. This means that there can be only a finite number of stars accessible to observation. In a universe with an infinite population of stars, any direction of view would end up on a star, so that the sky would appear bright all over. In a finite but static universe, stars would attract each other by gravity and would move towards each other first slowly, then more rapidly; such a universe would be unstable and contract. The other possibility, surprisingly not appreciated by Olbers or others in the last century, is of a universe of stars moving apart from each other – that is, the expanding universe we have described.

29

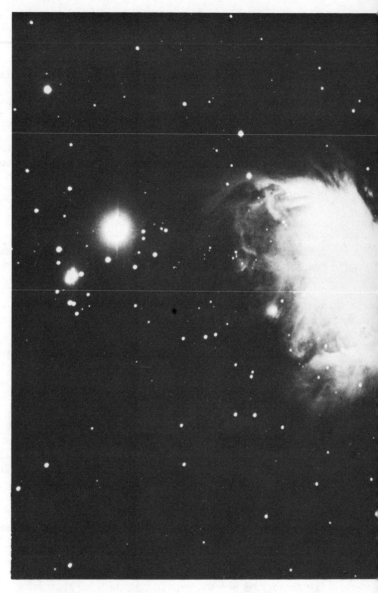

Fig. 2.9 *The Great Nebula in Orion where new stars are in*

the process of being formed (Royal Greenwich Observatory).

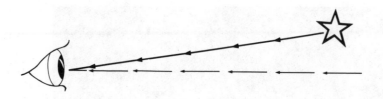

Fig. 2.10 *Olbers' paradox – the sky between the stars is dark at night. Some lines of sight* b *do not end up on stars (as does line of sight* a) *or other bright objects.*

Other evidence for expansion came with the classic work of Hubble in the 1920s, from observations of what is known as the red shift. Light from faint and therefore distant stars or galaxies is much redder in colour than that from comparatively nearby stars. This shift to the longer wavelengths at the red end of the spectrum occurs because these stars are moving rapidly away from us – in exactly the same way as the pitch of sound lowers (or the wavelength of sound increases) as a train, aircraft or car passes by us and recedes – the Doppler shift. The more distant the galaxy is from us, the more rapidly it is moving away. The most distant galaxies we can observe are moving away from us at an incredible 200,000 kilometres per second, over half the speed of light.

A third piece of evidence, crucial in the establishment of the Big Bang theory, came in 1965 when radio engineers at the Bell Laboratories in the USA demonstrated the presence of background radio emission from out in space. This background radiation which pervades all space is just what is left of the radiation present in such large quantities at the universe's beginning – a faint echo of the Big Bang. As the universe has expanded, so has the radiation, cooling from the billion degrees or so a minute after the Big Bang to the 3° absolute which is now the temperature of intergalactic space.

The fourth piece of evidence comes from the relative abundance of hydrogen and helium in the universe, a quantity which astronomers can easily measure. As we have mentioned, three minutes after the Big Bang, conditions of density

32

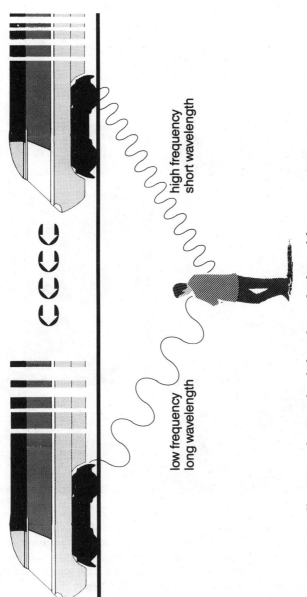

high frequency
short wavelength

low frequency
long wavelength

Fig. 2.11 *Illustrating the Doppler shift in the pitch of sound from a moving train.*

33

Fig. 2.12 *The James Clerk Maxwell Telescope – a 15-m-diameter radio telescope recently built and installed near the top of Mauna Kea in Hawaii. It can look for emissions from objects in the sky at millimetre wavelengths in the radio spectrum and investigate processes associated with galaxy and star formation and evolution (from R. W. Newport, Rutherford Appleton Laboratory).*

and temperature were right for helium nuclei to form, the ratio of helium to hydrogen being determined by the precise conditions of temperature and density which prevailed. From the average density of the present universe with its 3° absolute background radiation, the conditions which prevailed during the first few minutes of the universe's history can be estimated. Three minutes after the Big Bang the conditions were such that about one quarter of the nuclei formed would be helium, the rest being hydrogen – just the ratio which is now observed.

What about the process of nuclear synthesis taking place in stars and supernova explosions? How can we be sure that our understanding here is correct? Although the conditions prevailing in the interiors of stars cannot be reproduced at all easily in the laboratory, over the past twenty years nuclear physicists have employed many different types of nuclear accelerator in which collisions at high energy between different particles and nuclei can be observed. From the mass of

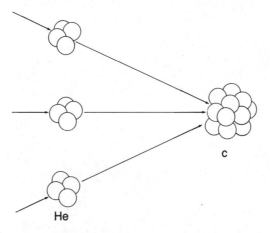

Fig. 2.13 *Three helium nuclei come together to make up the nucleus of a carbon atom. A critical resonance in the structure of the carbon nucleus assists this transformation ensuring that carbon is relatively abundant in the universe – of vital importance to the formation of life.*

35

Fig. 2.14 *The Space Telescope – a large optical telescope due
for launch in 1989 into orbit around the earth. Being outside
the disturbances of the atmosphere, it will enable astronomers*

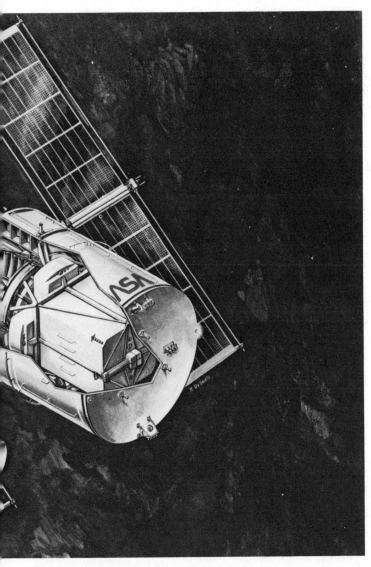

to view galaxies and other objects which are fainter and at greater distances than is possible to view from earth-bound telescopes (from European Space Agency).

data which has thereby been obtained, inferences have been drawn about the efficiency of different nuclear transformations. The relative abundances of different elements produced in cosmic processes estimated from this data fit in well with the observational evidence of the abundance of the elements in the universe.

So some large pieces of the jigsaw fit together rather well. Astronomers still argue extensively about many of the smaller pieces. They are also puzzled by the remarkable coincidences which seem to appear in the story. Conditions in the Big Bang were just right for the universe to expand at a rate such that galaxies and stars had time to evolve – time which was needed for them to produce the range of ninety-two elements in the abundances required for life on planet Earth to be able to exist in all the variety and complexity that we know.

Further, it was necessary for the Big Bang to begin extremely uniformly and yet with sufficient local variations for the formation of galaxies to be possible. The whole sequence has occurred with extremely high precision and exact timing.

There is also the question of what happened before the Big Bang – if indeed that is a meaningful question for us. It may well be that in due course an overall theory – physicists are fond of talking of GUTs (Grand Unified Theories) – will be developed which embraces the Big Bang itself and which ties all the seeming coincidences together, showing them to be simple consequences of the basic structure of things. Alternatively, some suggest that there may be a very large number of universes, all starting off differently. We happen to be in the one that has produced conditions right for us. Clearly, since there is a planet Earth with human inhabitants, the universe had to be such that the materials and the environment were there to provide for man's existence.

---

[1]S. Weinberg, *The First Three Minutes* (André Deutsch, 1977).
[2]A popular account of the history of modern cosmology can be found in T. Ferris, *The Red Limit* (Corgi, 1979).

Chapter Three

A GOD WHO HIDES

Truly, you are a God who hides himself. (Isaiah 45:15)

Very briefly in chapter 2 we outlined the story of the universe from the moment of the Big Bang to the formation of planet Earth. A story equally remarkable (if not more so) could be told about the structure of life, with the many complex and interdependent molecules which make up even the simplest living cell. The physics which describes the basic building blocks of matter and which is investigated in big particle accelerators or in careful laboratory experiments is the same physics that applies in the farthest galaxies, billions of years removed in time, and the same physics that applies in the intricate molecular forms in living matter.

Man formed of complex living machinery can look out on distant skies and galaxies. As we saw in the last chapter, the ninety-two elements from which he, man, is made and which he utilizes for his varied industry have been formed in stellar furnaces and exploding supernovae and then recycled as stars and galaxies die and are reborn. It can be argued that, for man to exist, the whole of the universe is needed.[1] It needs to be old enough (and therefore large enough) for at least one generation of stars to have evolved and died to produce the heavier elements, and then for there to be time too for a second-generation star like our sun to form with its system of planets. Other stars may well possess planetary systems including planets like Earth, inhabited by cognizant beings; although our chances of knowing of this or of making contact with them seem remote in the extreme. We might also speculate that other universes similar to our own could exist, although there is probably no way even in principle by which we could know of their existence. Here on earth we are seemingly complete and confined, a tiny (but, we may argue, not unimportant) speck in a vast universe operating under laws of the highest precision. The aim of the physicist is to explain all, and that his description should be complete and self-contained.

Much detail in the physical picture remains to be filled in. At the fundamental level, the search for a Grand Unified Theory continues, with the object of tying together the elementary forces and particles into one all-embracing scheme. At other levels, the power of the physical and mathematical tools available to the scientist is continually demonstrated as phenomena in the physics laboratory, in the cosmos or in living matter move from the territory of the unknown to that of the directly perceived. Although much remains to be done and no end is yet in sight for the scientific enterprise, there seems no reason to doubt that in principle the scientific picture can be complete.

Where, if anywhere, in this description is there room for God? And for what sort of God? Russian cosmonauts have visited outer space and reported that they cannot find him. Should we, in any case, have any expectation of coming across God in a spaceship or being able to bring him into the focus of an earthbound telescope?

Early scientists would invoke God as the explanation of phenomena their science could not explain. The most frequently quoted example concerns Isaac Newton who, having discovered that the law of gravity accounted neatly for the rotation of the moon about the Earth, found difficulty in explaining the spinning of Earth on its axis. He wrote to the Master of his Cambridge college, Trinity, 'The diurnal rotation of the planets could not be derived from gravity but required a divine arm to impress it on them.' Because 'God' provided a convenient explanation for the gaps in man's understanding, he came to be described as 'the God of the gaps'.

As science has advanced and provided more and more 'explanations', the need to account for the unexplained in terms of God has continued to recede. Although we commonly continue to describe as 'acts of God' some unpredictable natural phenomena – the lightning strike, the volcano or the earthquake – we feel that their unpredictability is merely a consequence of our scientific ignorance. They are probably not unpredictable in principle. Even in the medical sphere, it is often thought that God is less in demand as the healer. As a friend of mine commented, 'Penicillin is worth an awful lot of prayer!'

Now everything seems at least in principle amenable to scientific investigation and 'explanation'; although some suggest that perhaps God might at least be considered responsible for having pulled the trigger for the Big Bang at the beginning of time, thereafter leaving the universe to run itself. However, it already seems likely that, as science progresses, some scientific 'explanation' of even that initial singularity may eventually be provided.

Where then is there room for God? If God is confined to those parts of the universe science has not yet reached, those supporting belief in God are supporting a lost cause. By searching for God within the universe, we are looking in too confined a context. A vantage-point is needed from which a bigger view can be taken. Let us try to step outside the universe and think about it as a whole.

The idea of God as the 'great designer' is an old one. The eighteenth-century philosopher William Paley developed the idea into his famous proof of the existence of God, which supposes that a man finds a watch in a forest. The watch is ticking away, its mechanisms all moving together. The finder would immediately argue that someone must have designed and made such an elaborate device. *A fortiori*, said Paley, the existence of the universe demands that there be a designer. Although we may not accept the Paley 'proof' as a logical argument, it is nevertheless a helpful analogy which we can pursue further.[2]

Having found the watch, how do we find the watchmaker? Clearly he cannot be found within the mechanism of the watch itself. However, supposing that the watchmaker maintains an interest in his watch, he might be expected to appear from time to time to wind it up, to adjust it to the right time or to clean and maintain it. Unless, that is, the watch is sufficiently elaborate and accurate not to need any such attention. Even early watch and clock makers aimed at automatic adjust-

Fig. 3.1 The Voyager spacecraft which has explored the outer solar system, Jupiter, Saturn, Uranus and Neptune. Such spacecraft, which have to continue to function over many years of travelling through space, must be constructed to be very reliable and fault-tolerant (from NASA).

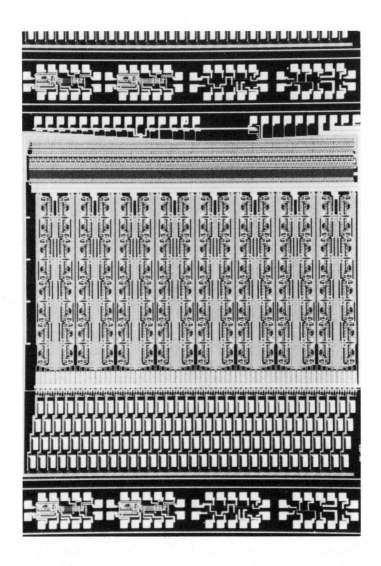

Fig. 3.2 *A complex computer circuit. The layout of an integrated circuit chip (dimensions 6.4×6.4 mm) for processing the information from a high-energy physics experiment (courtesy of Rutherford Appleton Laboratory Technology Division).*

ment and maintenance. For instance, in the palace of Versailles, there is a clock dating from the eighteenth century which records not only the minute and the hour, but the day, month, year and phase of the moon, with mechanisms that take account of all the necessary adjustments for leap years and the like for many centuries ahead.

The better constructed and more complete the timepiece, the less need there is for the watchmaker to attend for maintenance or adjustment. The more elaborate and perfect the watch, the less likely therefore are we to meet the watchmaker. Indeed, the ideal timepiece is one which needs no attention at all, in which case the watchmaker need never appear.

I have spent many years designing and building scientific equipment for artificial satellites. Such equipment needs to be extremely robust and reliable. Hands-on maintenance is not normally possible; the cost of visits by the space shuttle is prohibitive. In his design, the space-equipment designer needs to consider all possibilities of failure, building in redundancy and including back-up systems wherever failure is likely. Automatic means of correcting for failure need to be provided. The technical description for such equipment is 'fault-tolerant'.

Modern computing equipment has a similar requirement. The circuitry built into a large silicon chip can be so complex that complete testing of it is not possible. Some failures would be difficult if not impossible to identify. A degree of fault tolerance is therefore essential. With fault-tolerant equipment that is properly designed, there is no need to appeal to the designer for maintenance or adjustment. It should continue to function according to its specification for as long as its design life. After that, it can be discarded and replaced.

In a very different sphere, the human body is an excellent example of fault-tolerant equipment. Built into the body's biochemistry are elaborate mechanisms to combat or eliminate disease. A substantial degree of redundancy has been provided in the brain and other parts of the body so that the body can still function even if large parts are damaged or removed. The human body also possesses a remarkable ability to adapt to a changing environment.

The biological sphere contains many examples of fault

44

tolerance – a theme which is expanded by Richard Dawkins in his book *The Blind Watchmaker*,[3] in which he expounds what he believes to be the great capability of the mechanism of natural selection for adaptation and adjustment. He argues that nothing further is required to explain the development of the living world, and that, having found the mechanism, there is no need to invoke the existence of a designer; the mechanism of natural selection can be considered as 'the blind watchmaker'.

The fact that we understand some of the mechanisms of the working of the universe does not, of course, in any sense preclude the existence of a designer any more than the possession of insight into the processes by which a watch has been put together, however automatic these processes may appear, implies that there can be no watchmaker. Many scientists, including myself, feel that, even though no logical argument can be provided leading from the universe to a designer, the evidence tends strongly to demand the existence of an intelligent being behind it all. Other scientists, although not dismissing the possibility, do not feel the need to be pressed to such a conclusion.

Our attitude in this debate will be influenced a great deal by how much we can know about the designer. Let us therefore ask the question, 'Supposing there is a designer God behind the universe, what can the universe tell us about him?'

In this connection, another definition of God which is relevant and helpful is that introduced by Anselm of Canterbury in the eleventh century in connection with another of the classic 'proofs' of God's existence. Anselm called God the 'greatest conceivable being', and argued on logical grounds that because what exists in reality is greater than that which exists only in thought, it is necessary, if God is the greatest conceivable being, for him to exist in reality as well as in thought.

The validity of Anselm's logic has been debated by philosophers for centuries. We may not find it very convincing today. Nor are we looking for a formal proof of God's existence. However, the definition helps us to expand further our view of God.

If we are to look upon God as the greatest conceivable being, he must be not only the great designer but the greatest possible designer. He would design a universe entirely reliable, precise, without need for continuous adjustment and with a high degree of fault tolerance. We would not therefore expect God to have to appear to push the system back into line or to correct for unforeseen errors. It is not surprising that the universe appears so complete and self-contained.

The watch we found, automatic and accurate though it is, still needs a source of energy to keep it going. This could be an extremely long-life battery, the movement of the wearer's wrist, or light from the sun impinging on solar cells and so generating electrical power. Whatever it is, the source of energy is an essential part of the design and therefore the concern of the designer. How is the universe kept in being? It is sometimes said that it maintains itself, whatever that may mean. In which case, if there is a 'great designer', he must have provided for self-maintenance. I believe, however, it makes more sense to think of God not only as the great designer but as the source of that which keeps the universe in existence. God is then the great sustainer through whose continued moment-by-moment activity the whole show of the universe continues.

I have been arguing that if we are going to think of God at all, it only makes sense to have the biggest possible view of him. As we shall see in a later chapter, the picture of God as designer, creator and sustainer of the universe is one which accords well with that found in Hebrew thought in the Old Testament and the beginning of Christian thought in the New Testament. For Hebrews and Christians alike, learning about the universe (even the very limited universe they knew) has been a first step to learning about God.

---

[1]See, for instance, M. Rees, 'The Anthropic Universe', *New Scientist*, 6 August 1981, pp. 44–47.

[2]H. Montefiore, in *The Probability of God* (SCM, 1985), develops the idea of God as the great designer.

[3]Longmans, 1986.

Chapter Four

SPACE AND TIME

Nothing puzzles me more than time and space; and yet nothing troubles me less, as I never think about them. (Charles Lamb)

In trying to think further about God's relationship to the universe, it will help if we understand something of the strong link between space and time which underlies modern science. The scientist speaks of space-time, putting space and time together as one word; he also speaks of time as the fourth dimension. The purpose of this chapter is to explain what is meant by these expressions.

In the second chapter, when describing events in the universe, we noticed that, because of the finite speed of light, events at a distance away are necessarily seen at a certain time in the past. The Crab Nebula explosion to which we referred in that chapter was seen by earthbound astronomers 5,000 years after it happened; the recent supernova explosion in the Larger Magellanic Cloud, to which we also referred in chapter 2, was observed on 23 February 1987, some 155,000 years after it happened (fig. 4.1). Events at the edge of the observable universe are seen as they were perhaps 10 billion years ago. In our description of the universe, space and time are inextricably linked.

We speak of the dimensions of space and time. What do we mean by these? The three dimensions of space are familiar enough to us, often conveniently chosen as north–south, east–west and up–down. But they do not have to be that way. For instance, if I make a plan of my house, it is convenient to

Fig. 4.1 *The supernova (the bright star on the right of the picture) in the Larger Magellanic Cloud seen four days after it exploded on 23 February 1987. It is 155,000 light years away*

and 10,000 times brighter than any other object in the Larger Magellanic Cloud (photograph courtesy of the European Southern Observatory).

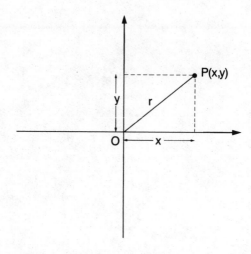

Fig. 4.2 *A diagram in two dimensions where any point* P *has two coordinates* x *and* y. *Pythagoras' theorem states that the distance* r *from the origin* O *to the point* P *is given by* $r^2 = x^2 + y^2$.

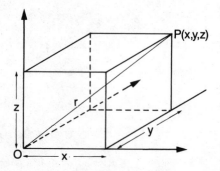

Fig. 4.3 *A diagram in three dimensions where any point* P *has three coordinates* x, y *and* z. *The extension of Pythagoras' theorem to three dimensions states that the distance* r *from the origin* O *to* P *is given by* $r^2 = x^2 + y^2 + z^2$. *The distance* r *is not, of course, dependent on the choice of the directions of the axes of* x, y *and* z.

choose the directions of two axes of the plan to be along the front of the house and along its side. But I could, if I chose, make other selections of axes. The choice of axes make no difference to the shape of the rooms or the width of the doors. The third dimension can be added, and elevations drawn of various parts of the structure of the house. For these it is normally convenient for one dimension to be horizontal and the other vertical. However, in drawing the roof structure, other axes may be selected to bring out particular features of the construction. If the drawings have been done correctly, the same answer will be obtained for any particular dimension within the house, whatever the drawing from which that dimension is deduced.

On Earth it is usual to draw plans in the horizontal dimensions and elevations including a vertical dimension. However, out in space in a spaceship, horizontal and vertical cease to have meaning; there is no up and down. For engineering drawings in space, three axes at right angles to each other would still be chosen. The choice would be made so as to be as convenient as possible, otherwise it would be an arbitary choice. In the universe as a whole there is no preferred dimension or direction.

How can time be brought in as a dimension to link with the three dimensions of space? Simple space-time plots are in fact quite familiar to us. We can, for instance, plot the train journeys covering, say, the 63 miles between Oxford and London. The diagram (fig. 4.5) includes all times between 0900 and 1500 hours and all positions on the railway line. Point X is the time at which I arrived at Reading station waiting for the train which I boarded at Y, arriving at Oxford at Z. Having drawn a diagram of one dimension of space with one of time, it is not difficult to imagine a three-dimensional model having two dimensions of space and one of time. The model could, for example, represent the horizontal position and the altitude of aircraft flying in a direct line between, say, London and New York (fig. 4.6). The third dimension could be marked off in hours so that lines constructed within the model would denote the positions of aircraft on that route as functions of time. Any point within the model would be described by three coordinates, two of space, say y (horizontal position) and z (altitude) and one of time, t. Following on from this, it is now

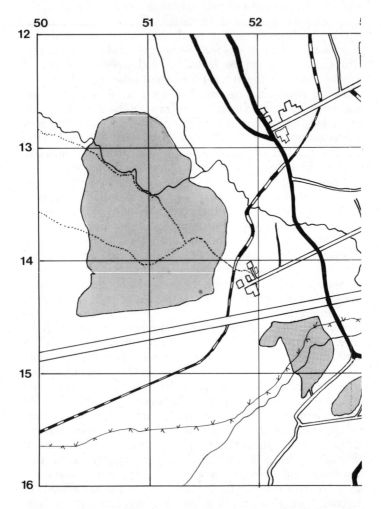

Fig. 4.4 *Any position on an Ordnance Survey map can be described by three numbers – two being the coordinates in the British National Grid Reference System, the third being the height above sea level. Because the Grid Reference System is a square grid – large squares are 100 km on a side – put down on the spherical surface of the earth, the sides of any square in the*

52

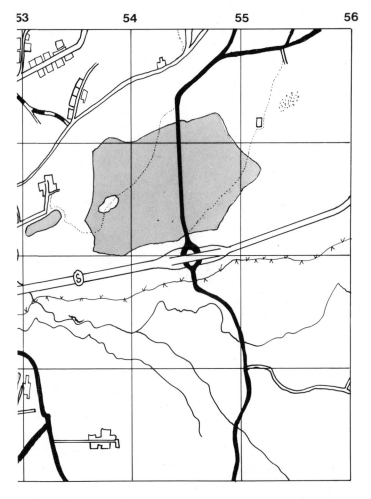

grid are not precisely N–S or E–W. A key on each map explains how far 'grid north' is from 'true north'. The information provided can also relate the grid reference to latitude and longitude. The choice of grid pattern or its orientation, of course, makes no difference to the distances between locations on the map.

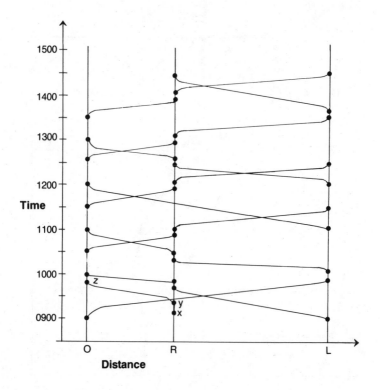

Fig. 4.5 *Distance–time plot of trains travelling between Oxford (O), Reading (R) and London (L) between 0900 and 1500 on weekdays.*

54

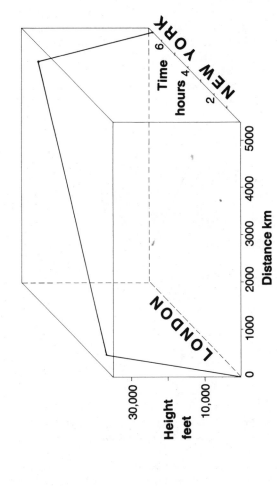

Fig. 4.6 A distance–height–time plot of an aircraft flight from London to New York, 5500 km and 7 hours flying time away. Much of the flight is spent at the cruising altitude of 35,000 ft.

perhaps not too difficult to imagine a four-dimensional model in which points are described by three space coordinates x, y and z, and one time coordinate t. Although we cannot, of course, actually construct a model, we can write down the rules for its geometry.

In the models we have mentioned so far, time and space are distinct. A big leap in thought was made by the Russian physicist Minkowski in 1908, when he put space and time together in a geometrical model in such a way that some of the distinction between space and time is removed. In Minkowski's geometry, time is made to look like space (fig. 4.7).

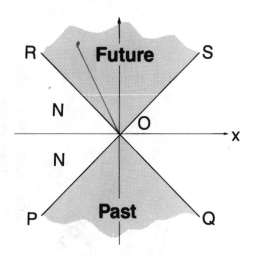

Fig. 4.7 *Space-time geometry*
The above is a space-time diagram (appropriate to a particular observer) with one space dimension x *and a time dimension in units of* ct *where* c *is the velocity of light. The point* O *at the origin is a particular position at the present time. Light beams from* O *can travel into the future along the diagonal lines* OR *and* OS. *Light beams from the past arrive at* O *along the diagonal lines* PO *and* QO. *Because no information can travel faster than light, only events in the past in the shaded area can*

communicate with or influence an event at O. *Similarly an event at* O *can influence only events in the future that lie within the shaded area. Events at points such as* N *cannot communicate with* O.

By analogy with distances in three-dimensional space which are independent of the choice of axes, 'distances' between events in the four-dimensional model can be defined. In the Minkowski space-time geometry, events are thought of as linked by light beams such that the 'distance' r *of an event* x, t *(taking, to begin with, just one space dimension) from the origin* O *is given by* $r^2 = x^2 - c^2 t^2$.

Because of the presence of the minus sign before $c^2 t^2$ *the distance* r *cannot be represented by lengths on a diagram. Note for instance (see diagram) that the value of* r *remains constant along the diagonal lines (i.e. along the paths taken by light particles).*

Extending the expression to include three space dimensions we have $r^2 = x^2 + y^2 + z^2 - c^2 t^2$.

A more extended semi-popular exposition of space-time geometry can be found in Space and Time in the Modern Universe *by P. C. W. Davies (Cambridge University Press, 1977) and* Extra-Galactic Adventure *by J. Heidmann (Cambridge University Press, 1982).*

By analogy to distances in three-dimensional space which are independent of the particular choice of axes, distances between events (defined in the caption to fig. 4.7) in the four-dimensional model are independent of the choice of space or time axes or of choice of origin. The presence of the minus sign in the equation for 'distance' means that the time coordinate t is in units of ict (c is the velocity of light) where i is the square root of minus one. Mathematicians call quantities multiplied by i 'imaginary' quantities. In popular jargon, therefore, an element of mystery is introduced into this fourth dimension with its 'imaginary' character. What Minkowski did, however, was to set up the rules for making a geometrical model of space-time such that in the model there is no preferred direction and no preferred time.

Minkowski's space-time geometry turned out to be an

effective way of describing Einstein's theory of relativity. In this theory, Einstein in 1905 elaborated the simple principles that there is no preferred velocity and that there can be no velocity greater than the velocity of light, *c*. These principles, which are rather simply embodied in the Minkowski geometry, lead to many results which are fundamental in modern physics. The most important and well known of these is the equivalence of mass and energy; mass can be destroyed and energy released (fig. 4.8). Atomic bombs and nuclear power

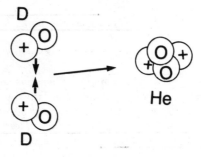

Fig. 4.8 *Two nuclei of deuterium D (heavy hydrogen) fuse to form a helium (He) nucleus. Mass m is lost and energy E released in the process which is the source of energy for the shining of the sun. According to the formula* $E=mc^2$, *which is a result of Einstein's theory of relativity, every gram of lost mass releases 30 million kilowatt hours of energy.*

stations are familiar earthbound demonstrations of this equivalence. The shining of the sun is the nearest extraterrestrial example of mass-to-energy conversion.

The success of the model which gives space-time a four-dimensional structure has led cosmologists to think of geometrical models of the universe. The most widely publicized of these is one in which the universe is a three-dimensional 'spherical' surface embedded in the four-

Fig. 4.9 *Image of the outer layers of the sun's atmosphere taken from a coronograph on the Solar Maximum Mission spacecraft. The sun's energy is derived from mass to energy transformation which occurs in nuclear reactions (from NASA).*

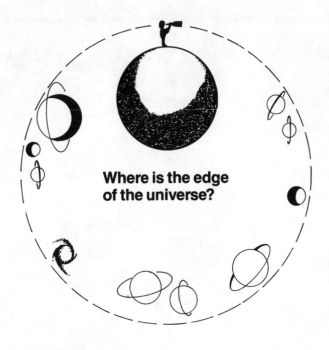

Fig. 4.10 *Where is the edge of the universe?*

dimensional space-time continuum. That sounds rather a mouthful, but a good idea of what is meant can be acquired by going back to the three-dimensional world we understand and looking at the surface of the earth, which is a two-dimensional surface embedded in the three-dimensional world. It is a curved surface, although when we look at small parts of the earth such as our garden, or even the town we live in, it is so close to being flat that for all normal purposes we neglect its curvature. Because the earth is a sphere it is an unbounded surface; it has no edges.

Being in the universe, the model suggests, is like being on the surface of a large sphere. Locally (and by locally the cosmologist would mean on the scale of the solar system or even the galaxy), for nearly all purposes the curvature is too small to be noticed. But on a very large scale the whole universe is curved, such that if we travelled in a straight line far enough and fast enough we would arrive back at our starting-point (fig. 4.10). I say 'fast enough' because, as we saw in chapter 2, the universe is expanding and we would have to travel impossibly fast to beat the expansion. On the same model, the expansion of the universe is seen as the 'spherical' surface expanding just as a balloon expands as it is blown up (fig. 4.11). All parts of the surface expand uniformly; letters on the balloon, small and almost unreadable to begin with, become larger and larger but all in proportion as the balloon grows larger.

Fig. 4.11 *Expansion of the surface of a balloon which is being inflated is like the expansion of the universe. No matter which part of the pattern on the balloon is marked with a X, all the other parts of the pattern move away from it, the pieces furthest from the X moving the fastest.*

Whether the universe will continue expanding, or reach a limit and then contract, or whether in fact the space-time continuum of the universe is spherical rather than hyperbolic or some more complex shape, are matters still hotly in debate among cosmologists. The purpose of this chapter has been to emphasize the developments that have arisen from the fundamental but simple idea that time should be included as a fourth dimension. Linking space and time in this way has had a profound influence on our way of understanding the physical world, both on the atomic and the cosmological scales.

Chapter Five

A FIFTH DIMENSION

Plato said that God geometrizes continually. (Plutarch)

In the last chapter we saw how much new understanding of the physical world flowed from the development of a geometrical model of the universe possessing four dimensions, three of space and one of time. We saw that not only is such a model helpful in aiding understanding but that it is also an essential tool in predicting the behaviour of physical systems.

Scientists are continually inventing models to help in their descriptions. For instance, the nucleus of an atom is extremely small, less than one millionth millionth (10^{-12}) of a centimetre in diameter, yet it contains many particles held together by a variety of forces. To assist in understanding the structure of the nucleus, scientists talk of the shell model (in which the particles are imagined to be arranged in shells similar to the way in which electrons are arranged in the structure of atoms) or of the liquid-drop model (in which the forces between particles are imagined to be similar to the forces between molecules in a drop of liquid). In a completely different field, scientists investigating the operation of the brain build models of aspects of its behaviour, relating them to the basic functions of a computer. Meteorologists build computer models of the atmospheric circulation for forecasting the future weather. Models of all kinds, thought models, computer models, scale models, practical models, are fundamental tools in scientific investigation.

Models are also part of the stock-in-trade of the theologian. Religious language constantly employs analogies or models.

For instance, Jesus in his parables introduced 'models' of the kingdom of heaven. The kingdom of heaven is like a man sowing seed, like a grain of mustard seed, like treasure hidden in a field, like a merchant in search of fine pearls, like a net gathering fish, and so on. Parables, metaphors, analogies and models abound in the New Testament and in the parlance of the modern preacher.

It is not therefore unreasonable to search for parables or models having a scientific base as we attempt to describe and gain some understanding of God's relation to the material universe. We have, for instance, great difficulty in knowing how to talk about where God is. As children we were perhaps taught that he is 'above the bright blue sky'. As adults, especially as adult scientists, we find that model unsatisfactory. Can science help by providing something better?

The crucial step in the space-time model described in the last chapter was the introduction of time as the fourth dimension. I want to suggest that we can think about God's position and relation to the universe as if he were present in an extra dimension. Let us see how such a model can help us in thinking about God. To aid in pursuing this model, we begin by trying to imagine life in a two-dimensional world. In the 1880s, an Oxford mathematician, Edwin Abbott, wrote a fascinating book entitled *Flatland*, in which he imagines a world having only two dimensions. The inhabitants of the world are confined to move on a plane, and indeed have no knowledge whatever of anything outside that plane. They experience north–south and east–west but cannot begin to conceive of up–down. For them the third dimension does not exist.

Abbott, presumably because he was a mathematician, imagines the two-dimensional world populated by beings whose outlines are mathematical figures: straight lines, triangles, squares, pentagons and so on to circles. One's class in Flatland society is determined according to the number of sides one possesses. The lowest class are the women with two sides – needle-shaped creatures with very sharp ends! The highest class in Flatland is the class of priests, who are circles possessing an infinite number of sides. Abbott describes in great detail how the different classes recognize and keep out of the way of each other; in fact the book was written as a satire on class.

64

"O day and night, but this is wondrous strange"

Ten Dim?

FLATLAND

Five Dimen *Eight D*

Seven *Six Dimen*

Nine

Four Dim en

| No Dimensions | A ROMANCE | One Dimension |
| POINTLAND | OF MANY DIMENSIONS | LINELAND |

By A Square

(Edwin A. Abbott)

Two Dimensions		Three Dimensions
☐		◻
FLATLAND		SPACELAND

My Study

The Pipe *MY BEDROOM*

My Sons

My Wife's Apartment *WOMEN'S DOOR*

THE HALL

MEN'S DOOR *My Wife* *My Daughter*

The Scullion *The Footman* *The Butler*

My Grandsons THE CELLAR

Policeman *Policeman*

"And therefore as a stranger give it welcome."

BASIL BLACKWELL . OXFORD

Price Nine Shillings and Sixpence net.

Fig. 5.1 *The cover of Abbott's book* Flatland.

Towards the end of Abbott's book, a sphere from the three-dimensional world of Spaceland appears, and attempts to explain to one of the two-dimensional inhabitants of Flatland what it means to possess another dimension and to be a sphere. The sphere passes through the plane of Flatland several times, appearing first as a point followed by a minute circle, a larger circle, a smaller circle again, and finally disappearing – a process completely inexplicable and magical to the Flatland inhabitants. The sphere then demonstrates that it can see into the interior of Flatland houses, rooms and cupboards without passing through the doors and windows – again utterly mysterious to the Flatlanders. Finally, the incredulous Flatlander is taken out of Flatland by the sphere from Spaceland and given a vision of the three-dimensional world. However, on returning to Flatland, he is completely unable, either through his attempts at descriptions or through mathematical analogy, to persuade any of the other inhabitants of Flatland to give any credence to his new-fangled ideas. To them, everything in Flatland is complete, there is nothing they perceive in their everyday life which cannot be described in two-dimensional terms. His seeing the circle and then not seeing the circle, and his conversation with this illusory being, were clearly hallucinations; things like that just did not happen in Flatland. To imagine any other than a two-dimensional framework for their existence seems completely unnecessary and impossible.

By way of analogy, let us imagine an extra dimension to the three of space and one of time with which we are familiar. We can call it the spiritual dimension which contains heaven, the dwelling of God and of other spiritual beings. Can this analogy help us in thinking about God and his relationship to the universe?

The sphere in Spaceland was normally outside Flatland yet could peer into all parts of the inside of Flatland; all events in Flatland were transparent to him. Further, he could enter and be present in Flatland anywhere he pleased. In a similar way, with the analogy of an extra dimension, we can imagine God in the spiritual dimension being outside the material universe yet being all-seeing and all-knowing regarding events within it, and having the ability to be present anywhere within it. In theological terms, God is both *transcendent*, that is, apart

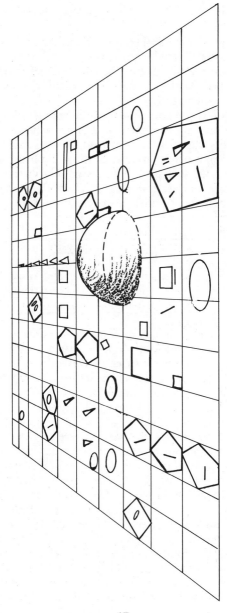

Fig. 5.2 *A sphere from Spaceland passing through Flatland.*

from the universe, and *immanent*, that is, present within the universe.

The analogy is helpful, therefore, in answering the question 'Where is God?' But why then have we called the spiritual dimension not the fourth but the fifth? This has been done deliberately because the analogy is also helpful in understanding God's relation to time.

Time is an integral component of the material universe. Its observation and measurement are intimately linked with material events, the movement of sand through the hour glass, the swinging of a pendulum, the beating of a heart, the rotation of the earth. It has meaning to us only in this physical context, and we have seen in chapter 4 how the dimensions of space and time link together in the creation of models of the structure of the universe.

Considering time as a dimension with similar characteristics to a dimension of space – as modern physics encourages us to do – enables us to think of God as outside time as well as outside space. Under this analogy, he can also enter the space-time world and appear within time. In our theological terminology, God is transcendent and immanent with respect to both space and time.

In our material existence we are so much creatures of time that the idea of being outside time seems a more difficult concept than being outside space. A very different analogy or model which is helpful is due to the writer Dorothy L. Sayers. She imagines God as the author of the human drama. The creator of a drama conceives the whole plot from beginning to end. Characters are introduced to take their place in the story as a whole. Even though the years covered by the drama are compressed to an hour or two in the theatre, to the audience watching the play on the stage, the plot unfolds gradually; all may be mysterious until the very last moment before the final curtain falls. The playwright, the creator of the drama, knows the whole story; he knows exactly the order of events within the play and how the final scene will emerge. He can be said to be 'outside' the time of the drama. There is also the possibility of the playwright, himself, being one of the players and entering in some sense into the time of the drama. Although helpful as an aid in consideration of the time dimension, such a model has its limitations. In particular, it does not do justice to the question of human freedom and responsibility; in the model we are all players on the stage, reading a prepared script. This problem we shall return to and look at more fully in chapter 10.

To add a further dimension in the way we have suggested is to add something very substantial. Just as three-dimensional objects are solid compared with two-dimensional objects, the model we have developed suggests that heaven, where God is, with its extra spiritual dimension, is a place of greater solidity than the material world we know. C. S. Lewis pursues a similar analogy in his book *The Great Divorce*,[1] in which he pictures inhabitants of hell arriving at the outskirts of heaven. Compared with the *solid* people from heaven who go to meet them, they appear as shadowy phantoms, transparent to the brightness of the place, pained by the roughness and sharpness of the solid objects around them, even of the blades of grass on which they walk.

A further point can be made regarding the way the dimensions interact with each other. A description of a world of two dimensions is completely contained within a three-dimensional description. Our model therefore suggests that the structure of heaven with its five dimensions contains the four dimensions which make up the material world; under this analogy, the space and time dimensions we know are part of the fabric of heaven. Rather than speaking of God and heaven as being *outside* space and time, it might be better to speak of them *transcending* space and time. In other words, although space and time are part of the structure of heaven, many of the limitations and constraints which they impose on events and movements in the four-dimensional space-time world are removed by the addition of a further dimension. Not only is there greater solidity in heaven, there is greater freedom.

In this chapter, we have been attempting to present models and analogies to assist us in thinking about God and where God is. An important question to ask is whether it is not improper or irreverent to think about God in this way. Are we trying to confine God by finding a slot for him, and calling the slot the fifth dimension? Such a danger is always there, although that is not, of course, our intention. On the contrary, what we have done is to look for analogies and models which do not confine God, it all being part of the mind-stretching

69

process of thinking about God as the greatest conceivable being. A helpful statement[2] regarding the limits and the use of pictures and analogies was made by the fourth-century saint, Hilary of Poitiers, who wrote, 'There can be no comparison between God and earthly things, but the weakness of our understanding forces us to seek certain images from a lower level to serve as pointers to things of a higher level. Hence every comparison is to be regarded as helpful to men rather than suited to God since it suggests rather than exhausts the meaning we seek.'

We should also realize that although the concepts expressed in the fifth-dimension analogy are ones which will be familiar to people with a modern scientific education, the ideas themselves are not new. In fact, they are present explicitly or implicitly in Hebrew-Christian thought from the earliest times. Many of the Jews and early Christians thought in somewhat spatial terms and imagined heaven, the dwelling-place of God, to be within a celestial sphere surrounding the Earth. They clearly found that a helpful picture. However, as expressed in the biblical writings, they did not try to confine God to a particular location. He was thought of as a spiritual being outside the universe he had created and also as present within it. God's transcendence and immanence pervade the whole of Scripture. Let us look at a few examples. Moses, when confronted by the burning bush,[3] realizes that God is not only in heaven but also present there within the bush. The name I AM by which God is called in that story gives hints, too, of his timelessness through his being ever present.

Of all the prophets, Isaiah is probably the one with the clearest view of God as creator and sustainer of the universe. God 'sits . . . above the circle of the earth . . . stretches out the heavens like a canopy, and spreads them out like a tent to live in.'[4] He created the stars, bringing out 'the starry host one by one.'[5] God 'inhabiteth eternity'; he dwells 'in the high and holy place, with him also that is of a contrite and humble spirit.'[6]

As a final example, moving to the period of the early church, we find Paul standing on Mars Hill, preaching to the Athenians and making use of ideas of the immanence of God present in contemporary Greek writings. He tells them that 'God who made the world and everything in it . . . does not

70

live in temples built by hands . . . he is not far from each one of us. "For in him we live and move and have our being." [7]

In this chapter we have suggested that, through the analogy of the fifth dimension, it is possible to think of God as transcending and yet within both space and time. The question may be asked, however, whether there is any point in attempting to stretch our thinking by these analogies. Are they just of academic interest, or are they of help, for instance, in getting to know God? I will argue in later chapters that although we are creatures limited by the four dimensions of space and time it is possible to have knowledge of God in the spiritual dimension, knowledge which is also relevant to our material existence.

---

[1]Fontana, 1972.
[2]Quoted by T. F. Torrance in *Space, Time and Incarnation* (Oxford University Press, 1969).
[3]Exodus 3.
[4]Isaiah 40:22.
[5]Isaiah 40:26.
[6]Isaiah 57:15 (Authorized Version).
[7]Acts 17:24, 27–28.

Chapter Six

A PERSONAL GOD

*The Son is the radiance of God's glory and the exact represen-
tation of his being.* (Hebrews 1:3)

In chapter 3, we thought of God as the great designer, whose
design of the machine we call the universe is so perfect and
complete that, even though we may think of God as the
sustainer of the whole, he is bound to appear elusive within
the complexity and grandeur of it all. The question might then
very reasonably be asked whether it matters if we believe God
is there or not. In chapter 5, the analogy was explored of a
fifth dimension where God is. We, however, are creatures
confined to our world of four dimensions and cannot readily
connect to the spiritual world of the fifth. We might therefore
reasonably ask whether the question of its existence is not a
merely academic one. The question of deciding whether God
runs the machine of the universe, or whether it runs itself,
might merit a certain amount of philosophical discussion; but
the conclusion we come to (if indeed we come to one) would
not have much impact either on our view of the universe or on
life as a whole. Thinking of God as a great and clever Force of
some kind may give us a mildly comfortable feeling but does
not really represent a lot of progress.

In chapter 3, God was described as the greatest conceivable
being. So far, in describing the relation of God to the universe,
we have done little more than imagine the greatest conceivable
engineer, who has devised a highly complex machine. God, if
he is to be the greatest in all respects, must also possess
personal qualities – qualities of self-awareness and the

73

potential to form relationships with other beings. Realizing this gives immediate point to our search. If we can know God in some personal way, if we can form a relationship with the great designer and sustainer of the universe, we are likely to be talking of something very big indeed. Some limited knowledge of another person can be gained from knowing about what he has done. But we can really get to know another person only if he communicates his personality to us through the variety of means of expression with which we as human beings are familiar. It follows that we can discover about God's personality only if *he* chooses to communicate with us. At the centre of the Christian faith is the belief that God has done just that, not in a way outside our normal experience – like the futile attempt of the sphere from Spaceland to communicate to the inhabitants of Flatland – but in the person of a unique human being, Jesus. It is hard to think of any other way in which communication of God's personal character could be arranged; personality can be conveyed only through personality. Even so, the incarnation, the technical term theologians employ for the coming of God into our human world, not surprisingly raises profound difficulties of comprehension – a problem to which we shall return in the next chapter.

It is not my purpose here to try to expound to any great extent the characteristics of the person of Jesus. I could add little to the many volumes which have already been written. Also, I am sure that the best way to begin to appreciate the character and personality of Jesus is to sit down with as fresh a mind as possible, and read, preferably in a modern translation, the accounts we have of his life, death and resurrection as presented by the four gospel writers. Since our interest here, however, is exploring helpful analogies which can assist in our search for connections between ourselves, the universe and God, it is worth looking briefly at some of the metaphors used by the New Testament writers and by Jesus himself to illustrate his character and his role in acting as a bridge between man and God.

The fourth gospel, the Gospel of John, begins with a telling metaphor that Jesus is the Word of God. It would immediately have struck chords with readers from both Jewish and Greek backgrounds.[1] To the Jew the word of the Lord was an extension of the divine personality, invested with the divine authority; for instance, 'By the word of the Lord were the heavens made.'[2] To the Greek, the Word (the Greek word *Logos*) represents God's self expression and includes the idea of the rational principle behind the universe. By introducing such a metaphor right at the start of his gospel, John establishes common ground with his readers from whatever background they may have come.

Also in John's Gospel, Jesus unequivocably identifies himself with God: 'Anyone who has seen me has seen the Father'[3] (meaning God the Father), replies Jesus to a question from one of his disciples about how they can see 'the Father'. The name I AM, which was one of the Jewish names for God (*cf.* chapter 5), was also taken to himself by Jesus. The Jewish leaders were puzzled that Jesus could claim personal knowledge of Abraham, who had lived so many centuries before. 'You have seen Abraham!' they jibed – to which Jesus replied, 'Before Abraham was born, I am.' The Jews took this remark of Jesus with its use of the divine name I AM as a claim to deity, and, because of what they saw as blasphemy, immediately took up stones to try to kill him.[4]

Further in the same gospel, seven down-to-earth metaphors are introduced, through which Jesus conveys the message about himself:

They are: 'I am the bread of life.' (6:35)
'I am the light of the world.' (9:5)
'I am the gate for the sheep.' (10:7)
'I am the good shepherd.' (10:11)
'I am the resurrection and the life.' (11:25)
'I am the way and the truth and the life.' (14:6)
'I am the true vine.' (15:1)

All of them emphasize man's basic need for a relationship with God. And they emphasize the reality of a life as real as, but more fundamental and more lasting than, our physical life. Variously called spiritual life, or eternal life, it is not something other-worldly that has no connection with our physical existence and everyday happenings. Rather, the point Jesus is constantly illustrating by his words and by his life is that it is only if proper connections are made to the spiritual world that life on earth with its attitudes, ambitions and relationships can properly be fulfilled.

The first and the last metaphors emphasize the basic truth that 'man does not live on bread alone'.[5] Just as there is a need to nourish physical life, so spiritual life too needs sustaining. The striking claim made by Jesus is that he is the source of this sustenance. We need to feed on the bread of life and to draw strength by being joined to the true vine. The central sacrament of the Christian church, the holy communion, in which bread is eaten and wine is drunk,[6] is simply and effectively a regular proclamation of our fundamental spiritual need and how it can be satisfied. More striking offers and claims from Jesus are contained in the other metaphors – illumination for our path in a dark world; guidance, leadership and protection from the Good Shepherd, and the guarantee of fulfilled life here and after death. Finally, there is the most comprehensive statement of them all: that Jesus claims to be 'the way, the truth and the life'.

At this point we may well ask: If we are to take these claims seriously, how can we connect so closely with someone who lived in Palestine nearly 2,000 years ago, so far removed from us in both time and space? We find that question addressed in the Gospel of John. Before Jesus left the world he promised someone he called the Counsellor to be with his followers to convince them of the reality of his presence and the truth of his teaching. This is the person we call the Holy Spirit.[7] In thinking about God in the spiritual dimension, Christians have found it helpful to think of God as a Trinity: God the Father who is outside the universe he created, God the Son who came and lived in the world but in a way which was limited in time and space, and God the Holy Spirit who is not so limited and who is God active and working in the world. It is he who helps[8] us in our struggle to get to know God.

In this getting to know God, we need to go to two main sources of information – the created universe on the one hand and the person Jesus on the other. The sixteenth-century reformers who had rediscovered the Bible talked of two books, the Bible and the book of nature,[9] as guides to faith.

A quotation from Francis Bacon reproduced by Charles Darwin on the flyleaf of *The Origin of Species* speaks of the book of God's Word and the book of God's works. From the book of nature we learn of God's greatness, grandeur and consistency, and the enormous scale on which he operates;

from the person of Jesus as presented in the book of the Bible we learn of his grace, love and purposes for human beings. Put these two views together and we begin, however dimly, to grasp the divine perspective. The God of our universe (and for all we know of other universes too) with its billion galaxies each with billions of stars in their different stages of evolution, the God of the living world with all its intimacy, variety and beauty, this same God is the one who has revealed himself through Jesus, who (using another metaphor or model) is described as God's Son.[10]

But now the question very naturally arises: what is the evidence for the picture we have presented? Is it not all a rather pious invention? Are we not creating a God in man's own image by extrapolation from our experience mixed with wishful thinking? Can we be sure of the existence of the spiritual dimension and of our perception of it?

There are three main lines of evidence: first, the historical record of the life, death and resurrection of Jesus; secondly, the universal awareness of a spiritual dimension; thirdly, individual personal experience. We shall briefly look at these in turn.

Nearly all the historical evidence we have is contained in the documents of the New Testament. These have been carefully studied by historians, archaeologists and experts in linguistics and are generally believed to be genuine.[11] The various accounts of the life, death and resurrection of Jesus reinforce each other in the main, but differ in points of detail in ways which are often typical of eyewitness accounts. Historical evidence on its own, however, can never be completely convincing. The records of the past are inevitably written with some particular perspective in mind. So far as we know, Jesus himself left no written record; the records we have were written by his disciples and followers quite deliberately from the standpoint of faith. Material was carefully selected and presented with the testimony of faith in mind.[12]

Clearly, the disciples were convinced. Those converted to the Christian faith within weeks of the resurrection were convinced. Many of them and many thousands since have been sufficiently convinced to suffer and die for their faith. We may well ask the question, 'Why should we be convinced?'

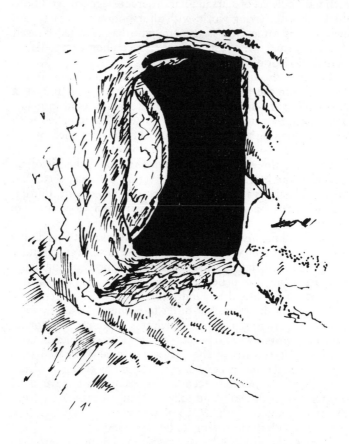

Fig. 6.1 *The empty tomb.*

Take, for instance, the central point of the resurrection. The main objections today arise from the feeling that an event so contrary to normal experience as a dead body coming alive again in resurrection requires far more than historical evidence for it to be accepted as fact.[13] People just do not rise from the dead and then appear and disappear in different places. However strong the historical evidence may be, it cannot be strong enough to compensate for the completely unusual nature of the events.

Were it just a matter of academic debate, we might be happy to remain sceptical. But then, if Jesus really is God who appeared in a body,[14] if he is God breaking into human history, if he is the way we get to know the greatest conceivable being, we look at the historical events in a different light and ask a different kind of question. We stop asking, 'How can I accept such unusual events as fact (even though the evidence for them may appear strong)?' and ask instead: 'Supposing they are fact, do they make sense?'

We shall return to the general question of miracles in chapter 10. Here we need to emphasize the particular point that, if Jesus is to convey the reality of eternal or spiritual life, if he is to demonstrate that he has triumphed over death and evil,[15] then his resurrrection is the key to his whole message. Let us go back to the two-dimensional world of Flatland and the attempt of the creature from Spaceland to convey the existence of the third dimension to the Flatland inhabitants. The Spaceland creature not only demonstrated that he could see inside the Flatland houses and the Flatland beings themselves and describe in detail what was going on, but he could also appear at will, albeit in a two-dimensional section, in any part of Flatland, and disappear at will. Jesus after his resurrection also appeared and disappeared at will,[16] demonstrating that he was no longer subject to the same limitations of space as he had been before. The resurrection of Jesus and his post-resurrection appearances support his statements and his message, namely that there is a spiritual dimension of fundamental importance which can be discovered through him as 'the way and the truth and the life'.

A consistency exists therefore between events in the life of Jesus (in particular his resurrection) and his teaching and declaration that he is the way to God. The complementary

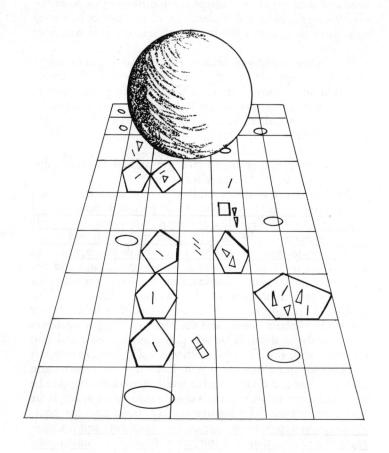

Fig. 6.2 *A sphere from Spaceland over Flatland.*

evidence that is now required is evidence that Jesus is alive today, not just in the sense that other historical figures live on, but in a much more active sense that we can each experience. The evidence for this is bound to be more subjective, though not entirely so. There is the general evidence that most human beings, from whatever part of the world and from the earliest times, have exhibited a fundamental belief in a divine being or beings, and in some sort of spiritual world. Then, more particularly so far as Christianity is concerned, there is the witness of millions of Christians over the years, whose lives have been transformed through contact with Jesus and who claim the reality of a continued, living relationship with him. Not that truth necessarily resides in the beliefs of large numbers or even a majority of people. But the experience of so many, especially the evidence of changed lives, cannot be lightly dismissed.

A strong, historical case and the witness of millions of Christians is, however, not going to convince me unless I have confirmation in my *personal* experience of the reality of a relationship with God. What sort of evidence must I look for? A few, like the apostle Paul,[17] come to see its reality as a blinding light, illuminating in a moment the whole detailed landscape of faith. For most of us, however, the road to belief is a much longer one. As with most human relationships, it is not so much love at first sight, but a growing awareness of the reality of God.

There is a gradual recognition that the Jesus I meet in the pages of the gospels is the one I meet as I attempt to communicate with God in prayer and the one I meet through the lives and conversations of others in the Christian community. It is through the practice of prayer that connections are made into the spiritual dimension. Through these connections, I meet in the same person the creator and sustainer of the universe and the one who died on a cross at human hands in order to rescue mankind from sin and failure.[18] As I realize his love not only for mankind in general but for me in particular, I am filled with feelings of gratitude and worship. Like a jigsaw puzzle, seemingly unconnected pieces of life, thought and experience begin to fit together. One piece on its own cannot be seen as necessarily part of the picture on the cover of the jigsaw box; when a number of pieces interlock and a picture begins to emerge, progress is being made.

Fig.. 6.3 *'Seemingly unconnected pieces of life, thought and experience begin to fit together.'*

My argument is, therefore, that there is a coherence between the historical evidence, the experience of the church over the centuries and my own personal experience. Given a position of faith, is the historical base of any importance? Some consider it of little relevance; faith, it is argued, can get along without its historical roots. For most Christians, however, myself included, the historical base is fundamental. Without it the whole structure falls to the ground. As Paul wrote very early in the church's history, 'If Christ has not been raised, our preaching is useless and so is your faith.'[19] The historical base and the experience of faith go along together; neither is sustainable without the other. All this is part of an integrated view which we shall attempt to explore further in later chapters.

[1]For an exposition of this and other metaphors in the Gospel of John, see William Temple, *Readings in St John's Gospel* (Macmillan, 1961).

[2]Psalm 33:6.

[3]John 14:9.

[4]John 8:56–59.

[5]Matthew 4:4.

[6]Luke 22:14–20; 1 Corinthians 11:23–29.

[7]John 14:16ff. and 16:7–8.

[8]William Barclay, in his two-volume commentary on *The Gospel of John* (St Andrew Press, 1975), suggests that the Holy Spirit's title usually translated as 'Counsellor' could be translated 'Helper'.

⟶ [9]The historical background to this idea is expounded in A. R. Peacocke, *Creation and the World of Science* (Oxford University Press, 1979).

[10]Matthew 3:17; 4:3; John 1:49; 3:16; Hebrews 1:2.

[11]See, for instance, P. Barnett, *Is the New Testament History?* (Hodder and Stoughton, 1986).

[12]John 20:30–31.

⟶ [13]For an evaluation of the importance of the resurrection, see G. E. Ladd, *I Believe in the Resurrection of Jesus* (Hodder and Stoughton, 1975).

[14]I Timothy 3:16.

[15]Matthew 9:6; John 3:16; 1 Corinthians 15:53–57; 2 Timothy 1:10; Hebrews 2:14.

[16]Luke 24; John 20 – 21.

[17]Acts 9.

[18]Romans 5:6–8; Colossians 1:20.

[19]1 Corinthians 15:14.

Chapter Seven

WAVES, PARTICLES AND INCARNATION

Two seemingly incompatible conceptions can each represent an aspect of the truth . . . they may serve in turn to represent the facts without ever entering into direct conflict. (L. V. de Broglie)

Three hundred years ago, the great scientist, Isaac Newton, thought a good deal about the nature of light. He carried out experiments, reflecting light from mirrors and refracting light through prisms, and came to the conclusion that all that he knew about light could be explained on the hypothesis that light consisted of a stream of particles. About the same time, a Dutch scientist, Christiaan Huyghens, also pursued some elegant optical experiments and postulated that light consisted of waves. However, so strong was Newton's reputation that the scientific world took little notice of Huyghen's ideas. A century later, Thomas Young reported the results of ingenious experiments demonstrating interference in light which had passed through a pair of slits. These strongly supported a wave theory of light. Even so, it was well into the nineteenth century before the wave theory was generally accepted, confirmation finally coming from the work of Clerk Maxwell, who elucidated theoretically the nature of the waves, showing that they were oscillations of coupled electric and magnetic fields.

So Newton was shown to have got it wrong and the wave theory reigned supreme – until the turn of the last century, when various new observations of the photoelectric effect and of the nature of black-body radiation were made which did

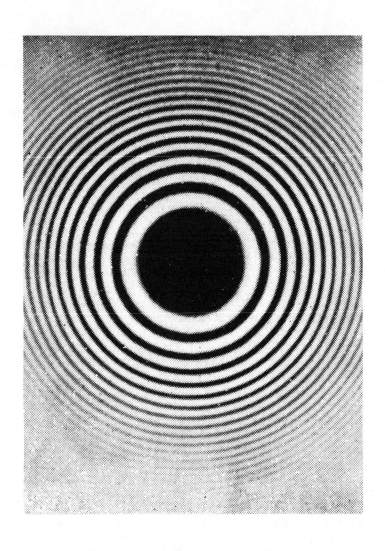

Fig. 7.1 *Patterns known as Newton's rings produced by the interference between two beams of light. Such patterns demonstrate the wave nature of light.*

not fit into the wave picture at all, and which suggested again that light consisted of particles. The quantum theory of light, which postulates that light consists of streams of very small particles or 'quanta', was put forward by Max Planck in 1900 to explain these new facts, so beginning a remarkable revolution which completely transformed nineteenth-century physics.

But the question whether light was to be described as waves or as particles persisted. Some properties, diffraction and interference for instance, were phenomena which could be associated only with waves. On the other hand, the way in which light interacts with atomic systems, the photoelectric effect and the existence of discrete spectral lines could be described only in quantum terms. Different circumstances therefore required different and seemingly contradictory descriptions.

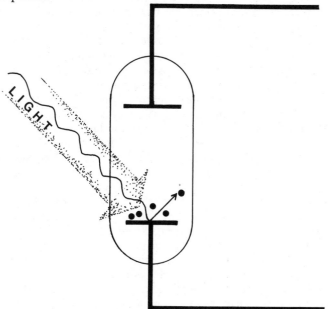

Fig. 7.2 *The photoelectric effect. Light incident on a suitable surface causes electrons to be emitted which allow an electric current to pass through a circuit containing the photocell. This effect is a demonstration of the particle nature of light.*

As the quantum theory developed, mathematical expression to some extent came to the rescue. Equations could be written down describing wave packets – groups of waves localized in space. Then came Schrödinger's equation describing the wave nature of the behaviour of matter, and finally the equation formulated by Paul Dirac, which fitted together electromagnetic radiation and quantum mechanics in an extremely elegant way. Dirac stuck firmly to his mathematical description, carefully avoiding the introduction of any pictorial model or mental picture of the phenomena described by his mathematical symbols; mental pictures, he explained, cannot be formed without introducing irrelevancies.

However, despite their limitations, mental pictures or physical models form an essential part of our scientific understanding. Although the duality inherent in the nature of light is accepted, and although we have a good idea under what circumstances the wave or the particle model can sensibly be applied, at the level of the model description, the paradox presented by the two contradictory models remains.

The dual description which is required of the nature of light can be employed as a parable of the paradox surrounding the person of Jesus. In the early centuries of the Christian era, a debate centred around theological schools in the two cities of Alexandria and Antioch, regarding how Jesus could be both God and man. The theologians at Alexandria emphasized particularly that Jesus was God. After all, they argued, his life clearly bore that out: his miracles, the authority with which he spoke. And did he not claim to be God, and was he not put to death on a charge of blasphemy?[1] But how he could be truly man as well was their problem. The theologian Athanasius, for instance, argued that what Jesus knew as God, he pretended not to know in so far as he was man.

At Antioch, on the other hand, they emphasized particularly that Jesus was man. He became tired and hungry; he wept over the death of his friend Lazarus.[2] And how could he save us or represent us if he were not genuinely human? How could he be God as well? So at Antioch, they spoke of 'two natures' – God and man each with his own person in a kind of intimate cooperation.

So the debate went on between them. In the year 451 at the Council of Chalcedon, the original version of the Nicene

Creed was reaffirmed in order to try to resolve the problems. It reads: 'We believe ... in one Lord Jesus Christ, the Son of God, begotten from the Father, only-begotten, that is, from the substance of the Father, God from God, light from light, true God from true God, begotten not made, of one substance with the Father, through Whom all things came into being, things in heaven and things on earth, Who because of us men and because of our salvation came down and became incarnate, becoming man ...'[3] We can imagine the length of the committee meetings that had produced that statement! But the debate did not stop at Chalcedon; it is, in fact, still going on.

Science and theology are full of paradoxical descriptions; the language of description has many limitations. Reality can never be reduced to one particular model. It is perhaps not surprising that we find difficulty when we try to find ways to describe the behaviour of elementary particles which make up an atom or an atomic nucleus. We should be even less surprised if we find difficulty in our attempts to describe the nature of the Deity. There are bound to be seeming contradictions and conflicts in the language we are forced to employ.

---

[1]Mark 14:64.
[2]John 11:35.
[3]Quoted in J. N. D. Kelly, *Early Christian Creeds* (Longmans, [2]1960), pp. 215–216.

Chapter Eight

DOGMA AND DOUBT

God forceth not a man to believe that which he cannot under-
stand. (John Wycliffe)

The motto of one of the world's oldest scientific academies,
the Royal Society of London, is *Nullius in Verba*. It can be
loosely translated, 'Take nothing for granted.' The success of
science depends on a thoroughly critical attitude on the part of
scientists. Scientific theories need to be given the most rigor-
ous examination. A new theory cannot be accepted until all
attempts to falsify it have failed.

In contrast, it is generally assumed that in the realm of faith,
much less rigour is required than in science. In matters of
religious belief, it is said there is no need to be so critical. So
much is this thought to be the case that the term 'theology' is
commonly used of a body of convictions which is not readily
subjected to criticism. If a commentator describes the debate
at a political party conference as a theological one, he means
that the discussion has become doctrinaire and of little
immediate practical relevance.

I want to suggest in this chapter, however, that despite these
popular views, there is, in fact, a lot of similarity between the
approach of science and the approach of faith.

First, let us look in more detail at the scientific approach. A
common view of science is that progress is made step by step
through relentless logic, each new achievement being based on
watertight argument. Attached to each could be the validation
Q.E.D.[1]

Although logic and argument have a great deal to do with

the scientific enterprise, however, it is not in practice as simple as that. In the first place, we cannot start from scratch; it is essential to build on what has been done already. In fact, a large proportion of the time required to tackle and solve a particular scientific problem will be spent in searching the vast scientific literature for papers of relevance and in extracting from the large body of existing scientific knowledge material which will help in the investigation. It is common to refer to the body of existing knowledge as the conventional or the accepted wisdom. In this body of knowledge some parts will inevitably be more firmly based than other parts. But, as with every activity, familiarity with what is known already is an essential prerequisite to further progress.

When it comes to a new scientific idea or theory, an important question is, How does it fit in? Does it accord with observations and experimental facts? Does it fit with our ideas of beauty, elegance and order in nature? Is it economical – in other words, can it describe a wide range of phenomena with the minimum of assumptions?

Take, for instance, the theory of relativity introduced by Albert Einstein in 1905, to which we referred in chapter 4. Only two assumptions were needed – that the laws of physics are the same within all frames (*i.e.* sets of coordinates), even though the frames may be moving relative to each other, and that the velocity of light is independent of the velocity of the source. As we saw in chapter 4, following from those simple but rather fundamental assumptions came predictions concerning the increase of mass with velocity and the equivalence of mass and energy. Elegance and economy were strong features of the new theory, and this readily persuaded many physicists, at the time, of its correctness. But there were others who were unconvinced. Where, they said, was the experimental proof? A complete unassailable channel of logical deduction could not be provided, and supporting experimental facts were scanty. It was not in fact until 1919, when, during a solar eclipse, it was demonstrated that light rays which pass near the sun are 'bent' (fig. 8.1) (one of the predictions of the theory), that much of the suspicion was removed. After that, Einstein still had to wait until 1922 for a Nobel Prize; even then it was primarily for his work on the photo-electric effect rather than for his theory of relativity.

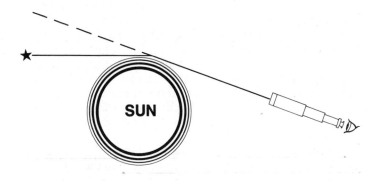

Fig. 8.1 *Observation in 1919, during a solar eclipse, of the bending of light rays from a star as they passed near the sun, provided an important confirmation of Einstein's theory of relativity. The deviation of the light ray is much exaggerated in the diagram – it amounts to only about one second of an arc.*

Now that it is relatively easy to carry out measurements on atomic particles which are readily accelerated to speeds near to that of light, many demonstrations of the accuracy of the predictions of relativity theory can be given. Nevertheless, if you were to ask a typical physics student to prove to you the truth of relativity, he would be unlikely to quote an individual example, at least until you pressed him. Rather, he would tell you that relativity is so fundamental, and that it influences so many parts of physics, that it is inconceivable that physics could exist without it.

Relativity is one of those fundamental theories of science whose great strength is that they have brought together ideas and observations from different parts of the subject and integrated them into new frameworks. The work of Clerk Maxwell is another classic example of the power of the integration of ideas. In 1864, he wrote down for the first time the equations which united light, electricity and magnetism into one theory. So the foundations were laid for applications in the electric power and communication industries upon which so much of our modern living depends.

It should not be imagined that these far-reaching theories of electromagnetism and relativity came completely out of the blue, unrelated to the knowledge which existed at the time. On the contrary, they built on many years, if not centuries, of previous work, of careful experiment and interpretation of experiments carried out by other scientists. Maxwell owed a great deal to the experiments and thought of Faraday. Einstein built on the work of many; Michelson and Morley, Fizeau, Lorentz and Fitzgerald are just some of the well-known names which can be mentioned.

Supposing, therefore, I say that I understand a theory such as the theory of relativity, what do I mean? In the light of the points I have made regarding the scientific approach, three elements can be delineated. First, I mean that I am familiar with the theory, I have read a lot of books about it, I have answered examination questions on it, I have immersed myself in what it is about. Secondly, it means that I know how to apply it. Given a physical situation concerning atomic particles, a nuclear power station or light signals in the cosmos, I can work out the result required. Thirdly, it means that I can see where the theory fits in with other parts of physics. This fitting in may not be complete. In fact, there are bound to be areas of difficulty, uncertainty or seeming contradiction. To that extent my understanding is bound to be limited. For it to be called 'understanding' at all, it is necessary for these three elements – familiarity, knowing how to apply it, and seeing where it fits in – to be present.

Science has developed by combining an acceptance and appraisal of the existing body of knowledge with a thoroughgoing, critical, questioning attitude. Pieces of the jigsaw are given; new pieces are continually being added and need to be put into place. Many groupings of pieces, all fitting together well, have already been established. Other groupings fit together less convincingly. New arrangements have to be tried; awkward pieces can be particularly helpful in providing clues as to possible new arrangements. Progress is made by applying an appropriate mixture of dogma and doubt.

Turning now to the theological scene, trying to make sense of the theological jigsaw is not, I believe, very dissimilar from making sense of the scientific one. The same elements which we have already mentioned are involved; namely, becoming

1 - becoming familiar with past efforts ...
2 - interpreting and applying observation and experiment ...
3 - integrating ideas ...

familiar with past efforts, interpreting and applying observation and experiment, and integrating ideas.

The basic data of the Christian faith have already been mentioned in chapter 6. We mentioned there the two 'books', the book of nature, and the book of the Bible. In the historical record of the Bible we read of the events surrounding the life of Jesus and his teaching (recorded in the gospels), together with the account of Jewish religion and culture in the Old Testament, which provides the necessary background to the events of the New Testament. We also have the interpretation of the life and ministry of Jesus by the first disciples, as recorded in the gospels themselves by the evangelists, the Acts of the Apostles and the letters of the New Testament. Then there is the continual record of Christian experience over the centuries, especially perhaps the testimony of individuals we know today.[2] With all that we need to become familiar.

What, then, about the second component of understanding, namely the application of Christian faith? Can I solve problems with it? Here we need to remember that faith is concerned with a personal relationship, not just with a set of facts. Not only my mind is involved, therefore, but all of my being. This personal factor in no way reduces the need for integrity, honesty and critical appraisal. It means necessarily, however, that the methods of application and the criteria by which the results are judged are bound to be more personal. Commitment is also necessary; not just the intellectual commitment present in the pursuit of scientific enterprise, but the commitment appropriate to the development of personal relationships.

The unique feature of the Christian faith is that the commitment which is required is not to a remote inaccessible God to whom we need in some way to work our way up, but to the person of Jesus, who is God come down to earth. It is as I commit myself to the person of Jesus, accepting that he can deal with my problems of guilt and moral failure, that my need for forgiveness and renewal is provided for.

Although this commitment can be said to be in the spiritual dimension, it is, of course, strongly linked to the ordinary dimensions of time and space. It is in the world of these ordinary dimensions that my choices are made and my failures occur. It is also in the world where my faith, if it means anything at all, is to be worked out in practice. Jesus, through his work in the world, showed how this can be done. It is not by accident, therefore, that Christians have often been the initiators of the provision of hospitals and orphanages, prison reform, famine relief and many other caring organizations and institutions.

The third component of understanding we have mentioned concerns the integration of ideas, or seeing things fit together. This is a particularly important component as far as religious faith is concerned, because religious thought must of necessity bring together strands from all aspects of our being and experience, weaving them together into a coherent whole. Knowledge of God, the greatest conceivable being, creator and supreme person, must provide us with a grand perspective from which we can see how the variety of our experience fits together. Earlier in the chapter, when considering some of the basic theories of physical science which enable large parts of the scientific jigsaw to be fitted together, we saw that scientists find it difficult to escape from the conviction that these theories are basically correct. So with the Christian faith many have found that the fitting together of the two revelations – of God as creator and sustainer on the one hand, and God revealed in Jesus on the other – is so convincing in their experience that they cannot escape the conviction that their faith is basically true. Not that all is seen clearly. Even more than in science, faith sees 'but a poor reflection as in a mirror',[3] but enough of the pattern is distinguished to provide not only conviction for the mind but a foundation for the whole of life.

The view that science and religion provide different and competing views of the world and of experience is a common one. It is also an unfortunate one. Belief in a being within whose creation is centred all our experience – scientific and religious – implies that truth must be seen as a whole. Centuries ago, the apostle John wrote: 'God is light; in him there is no darkness at all.' Much more recently, Albert Einstein remarked, 'Subtle is the Lord, but malicious he is not.' Einstein's life's work was to expound some of the subtlety. Christian experience also finds plenty of subtlety, but it is the subtlety of light and not of darkness.

[1] *Quod erat demonstrandum*, 'which was to be demonstrated'.

[2] See, for instance, M. E. Callen (ed.), *My Faith* (Marshall Pickering, 1986).

[3] 1 Corinthians 13:12.

[4] 1 John 1:5.

Chapter Nine

EXPLAINING THINGS AWAY

Only wholeness leads to clarity. (Friedrich von Schiller)

One of the most powerful tools available to the scientist is that of analysis or breaking things down into component parts. Examples of the process readily come to mind. Atomic physics, for instance, is all about the particles making up atoms; nuclear physics is all about the components of nuclei and the forces that hold them together. Molecular biology has developed from an understanding of the properties of molecules such as DNA which form the basis of living material. The science of chemistry is based on knowledge of the make-up and structure of the molecules forming different compounds and materials. Water is an excellent solvent, because the way in which the hydrogen atoms and the oxygen atom which make up the water molecule share the electrons gives the molecule an unusually large electric moment. The strength and the high refractive index of a diamond are due to the arrangement of electrons within the elementary diamond crystal. The blue colour of the sky is due to sunlight scattered preferentially at the shorter (blue) end of the spectrum by nitrogen and oxygen molecules in the atmosphere – and so on.

The analytical approach is also effective in less conventional scientific disciplines. The classical experiments of Pavlov on dogs, for instance, showed how the process of learning could be understood in terms of stimulus and response. Psychologists and behavioural scientists in their turn look at the way human behaviour is related to genetic inheritance and childhood environment and relationships.

Because analysis is such a powerful tool in science, it is often argued that it is all-powerful, and that, having described the component parts, the interactions between them and the way they fit together, there is nothing more to be said. The whole is then nothing but the sum of the parts.

So strong has the analytical instinct become, that the all-embracing power of analysis has become with some a tenet of belief. If only we knew enough, the reductionists[1] argue, everything could be reduced to elementary particles, and elementary forces. The whole of science and of human experience is then nothing but the sum of billions and billions of elementary parts. Reductionism, it appears, has the power to explain everything away.

The falsity of the reductionist approach is often illustrated by taking the example of a painting. A chemist would describe it in terms of various chemical substances distributed in certain patterns over the canvas; a physicist might go into the spectral properties of the pigments; other scientists might be able to determine its age or the origin of the materials of which it is made. None of these descriptions has any regard at all to the beauty or the artistic message. In no way has that message been 'explained away' or made superfluous by the detailed analytical descriptions. In using the illustration of the painting, descriptions taken from different disciplines have been involved, and it has been rather easy to show that weakness of the reductionist argument. It is also interesting to realize, and important to emphasize, that even if we confine ourselves to the realm of physical science, where the analytical approach is most effective, it is not all-powerful.

In the last chapter we saw that those who have been able to make the big leaps in scientific thought have taken a broad and integrated view. Progress is made both through exploration of the component parts and also through standing back and putting different ideas together. No particular description can be said to explain things away or preclude the possibility of another description. Frequently, if we imagine the whole to be made up just of the sum of the parts, important areas of science itself (not to mention other areas of experience and knowledge) will be missed or badly misunderstood.

Let me demonstrate this by taking a couple of examples from physics. The first example comes from quantum mechanics. According to quantum mechanics, the location and behaviour of elementary particles can never be completely defined; further, the act of measurement itself can influence the result of a measurement. We mentioned in chapter 1 principles involving probability and uncertainty which lie at the heart of quantum mechanics, but which physicists find it hard to get used to, and about which some, including Albert Einstein, have felt considerable unease. Despite this, however, physicists now almost universally accept that there are fundamental reasons why the quantum-mechanical description is not only correct but complete.

The quantum description can, however, come up with surprising results. A particularly telling example of the peculiarity of quantum-mechanical prediction is demonstrated by some recent experiments carried out by Professor Aspect and his colleagues at the Orsay Laboratory in Paris (illustrated in the accompanying diagrams, fig. 9.1). In these experiments, pairs of photons (*i.e.* light particles) emitted by calcium atoms are examined. Their behaviour cannot be explained by treating the photons as separate entities even though during the experiment they may be many metres apart without any possibility of influencing each other. Even though there are only two particles, the whole is not the sum of each separately.

The second example comes from thermodynamics. Supposing we were able to make a film of the movements of the individual molecules of two gases A and B just after the partition separating the gases had been removed (see fig 9.2). The film would show the molecules colliding with each other and with the container walls, all motions being entirely consistent with Newton's laws of motion. Gradually the two gases would diffuse throughout the container until they were completely mixed.

Suppose now the film were to be run backwards. Again the molecular motions would appear entirely consistent with Newton's laws. From the point of view of these laws applied to individual particles there is no way of deciding when the film is moving forwards and when it is moving backwards. However, we know, of course, that gas molecules do not move so as to separate the two kinds of molecules into the two ends of the container. There is a physical law which states that, in such circumstances, disorder tends to increase with time; physicists call it the law of increase of entropy. The process of

101

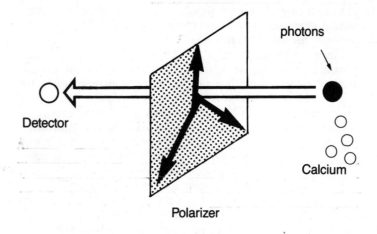

Fig. 9.1 *The Orsay experiment*[2]
Pairs of photons leave excited calcium atoms in opposite direc-
tions. Before arriving at detectors, which may be many metres
apart, they pass through polarization analysers whose purpose
is to measure the direction of polarization of the photons. These
can be oriented in any one of three directions 120° to each
other. The orientation of the analysers is changed continuously
and randomly. The signals from the detectors record the detec-
tion or non-detection of the photons. In accordance with the
predictions of quantum mechanics, it is found that (1) when-
ever the analysers are set in the same direction, the detectors
give identical signals, i.e. the photons are either both detected
or both blocked by the analysers, and (2) that whenever the

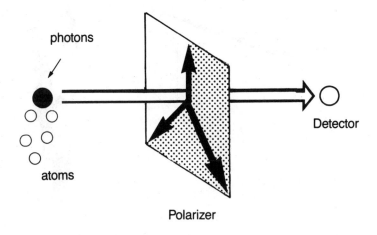

photons

atoms

Detector

Polarizer

analysers have different settings the detectors give identical signals only one quarter of the time.

The surprising feature of the prediction is that the probability of one of the photons passing through its analyser and being detected depends on the setting of the other analyser and the probability of the other photon being detected – the quantum-mechanical description treats both photons together throughout the whole event until they have both been detected, even though the detectors may be many metres apart. No set of instructions could be given to the photons as they leave the emitting atoms which could achieve the result. The formal proof of this is due to J. S. Bell in 1964 and is known as Bell's theorem.

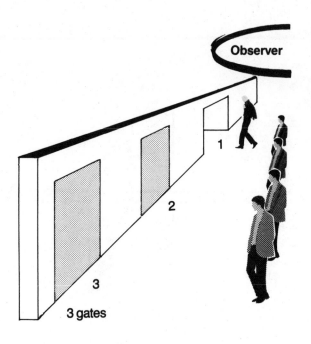

Observer

1

2

3

3 gates

Some readers may be unfamiliar with photons and polarizers and may find a description helpful in which the photons have been replaced by men and the polarization analysers by gates.

Men leave a central location in pairs, running in opposite directions towards two sets of three gates numbered respectively 1, 2 and 3. As each man arrives at the gates he will find only one gate open (chosen at random and constantly changing). He is unable to see his partner or the gates his partner is encountering as he makes his choice whether or not to go through the gate. The gate which is open is indicated to the observer on a display panel. Also indicated to the observer is whether each man goes through the open gate on his side. As a large number of pairs of men reach the gates, the observer notes the choices they make. He finds that (a) whenever the same-numbered gate is open simultaneously on the two sides, the two men make identical choices, i.e. they both go through or they both fail to go through, and (b) whenever different-

104

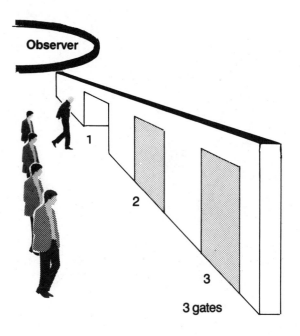

Observer

1

2

3

3 gates

numbered gates are open on the two sides, the two men are much more likely to make different choices (in fact, they make different choices three-quarters of the time).

Think for a few minutes about the findings of the observer. To achieve the first result, namely that identical choices are made by the men whenever the same numbered gate is open, it seems necessary that the men, before leaving the central location, should agree on a set of instructions as to the choices they will make dependent on which gate they find open. However, no set of instructions which can achieve the first result can also achieve the second, namely that when different gates are open the men will make the same choice only one quarter of the time.

To achieve the second result as well as the first, some collusion is necessary between the men when they reach the gates and discover which of the gates is open. No such collusion is allowed by the experiment.

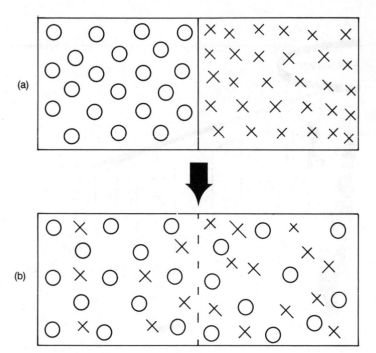

Fig. 9.2 *A container containing gases A (molecules labelled O) and B (molecules labelled X); (a) in separate compartments, and (b) after the partition between the compartments is removed.*

mixing of gases such as we have described is not reversible. In fact, it is one of the processes in physics which enable a direction to be given to time;[3] on viewing the two films, there is no difficulty in choosing the one in which time is moving forward.

We have seen, however, that the basic laws of motion are equally valid for time going forwards or backwards. By considering individual molecules, the concept of entropy or a direction to time will not emerge. To realize these, new ideas appropriate to the behaviour of the very large assemblies of molecules have to be introduced and new properties defined.

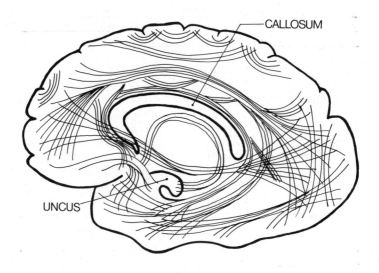

CALLOSUM

UNCUS

Fig. 9.3 *The human brain may be compared to a computer, although, with its 10 or 100 billion nerve cells, much larger and more complex than any computer yet constructed. This diagram shows the main lines of information flow between areas of the human brain (adapted from D. M. MacKay,* Brains, Machines and Persons, *Collins, 1980, p. 32).*

Even in basic physics the whole is more than the sum of the parts.

Turning now to a much more complicated system – the human brain – it is found to be helpful to compare the brain to a computer with its logic circuits (its hardware) and its programs (its software). A lot of progress in understanding the basic mechanisms of the brain can be made in this way. By analogy with our simple examples from physics, however, it would be surprising if the behaviour of a human brain could be reduced to be nothing more than the behaviour of a supercomputer. For instance, there are the properties of self-awareness and the ability to make free choices not possessed by computing hardware of human construction. We cannot argue that such properties are illusory because they do not appear in the behaviour of the individual computer elements (or neurons) which make up the brain, any more than we can argue that entropy is illusory because it cannot be found in the individual molecules of gas in a container. Neither can we argue that our free choices are not really free, without also denying our self-awareness.[4] We cannot look in particular areas of the brain for stuff called 'self-awareness' which has been added somewhere along the line in the brain's formation – although we shall, of course, look hard for links between properties of the whole, such as self-awareness and the micro-structure of individual parts.[5] None of the links we may find, however, will 'explain away' the property of the whole. Again it is important to realize that the whole is more than the sum of the parts.

What we have been doing in this chapter is giving examples of hierarchies of description.[6] Descriptions of objects or of events can be given at different levels, each employing the language and thought forms appropriate to its level. For instance, let us look at a sacramental meal, when a company of worshipping persons take bread and wine together. It can be described in terms of the basic biology of the growing process of the wheat or the grape, or in terms of the chemistry of yeast or fermentation through which wheat is turned into bread or grapes into wine, or perhaps in the framework of the agriculture and economy of cereal and wine production. It might also be described in terms of the social and behavioural patterns involved in human association and relationships and in the development of religious experience.

A further description might be given concerning the portrayal of the sacramental meal in art – by a Leonardo da Vinci or a Michelangelo – or its association with music over the centuries. Then, of course, there is its description in a Christian context as an aid to faith. There are connections between these different descriptions, but they are in no way mutually contradictory. If one description seems particularly adequate, the validity of a description at a different level is not called into question.

Fig. 9.4 'The Last Supper, after Leonardo da Vinci' by Anon. Milanese artist, 16th century. Trustees of the British Museum.

109

So, then, if we turn to the material and experience of faith, it is entirely proper that these should be analysed by the scientific tools at our disposal; such analysis can be helpful in leading to critical understanding. But it does not 'explain away' the reality of faith or of religious experience.

In the realm of faith, putting alongside each other different parts of the whole can be especially valuable. An underlying theme of this book has been the wholeness of God's revelation as seen both in the natural world and in his special revelation in Jesus. We shall return to this theme in chapter 12, where I believe we shall see again that the whole is greater than the sum of the parts.

---

[1]Believers in reductionism. Professor D. M. MacKay has nicknamed reductionism 'nothing-buttery'.

[2]Further details of the experiment can be found in N. D. Hermin, *Physics Today* (Phoenix, 1985), pp. 38–47.

[3]For more details see R. Feynmann, *The Character of Physical Law* (BBC Publications, 1965), chapter 5.

[4]Professor D. M. MacKay has argued (for instance in *Brains, Machines and Persons*, Collins, 1980) that freedom of choice is a logical consequence of self-awareness.

[5]A. R. Peacocke, in *God and the New Biology* (Dent, 1986), expounds this in some detail.

[6]A. R. Peacocke, *Creation and the World of Science* (Oxford University Press, 1978). See also R. Feynmann, *op. cit.*, p. 125.

get all of these

Chapter Ten

NATURAL OR SUPERNATURAL?

If they do not listen to Moses and the Prophets, they will not be convinced even if someone rises from the dead. (Jesus in Luke 16:31)

After quarks, believing in the virgin birth is a doddle. (Caption to cartoon in *New Scientist*)

From the earliest days of spectroscopy, it has been known that atoms, when heated in flames, give off radiations at characteristic wavelengths. We are all familiar, for instance, with the bright yellow emissions given off by sodium atoms when common salt is put into a flame or when sodium is present in a fluorescent lamp. In the early years of this century, one of the triumphs of the new quantum mechanics was its ability to provide an explanation for the existence of these discrete spectral lines. In 1913, Niels Bohr, the Danish physicist, set up a simple model of the atom in which the energy levels of an electron in an atom could be described by three numbers, called quantum numbers (fig. 10.1). The numbers are like the components of an address, for instance the town, the street and the house number, which enable any given home to be located. Spectral lines arise from electrons 'jumping' between these energy levels, emitting radiation in the process.

It turned out, however, that three numbers were not enough to describe some of the complications of observed spectra. In 1925, Wolfgang Pauli, an Austrian physicist, suggested a fourth quantum number, called the 'spin' quantum number. It seemed that the electron was spinning on its axis

111

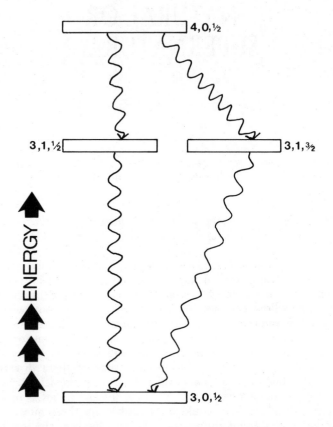

Fig. 10.1 *The first few energy levels of a sodium atom. The numbers by each level are the quantum numbers* n, l *and* j, *numbers which are employed to describe the electron configuration in each energy level. Transitions between the energy levels which result in the emission of light are shown. The yellow light at wavelengths of 0.5890 and 0.5896 micrometres characteristic of emission from sodium atoms results from transitions from the first two excited levels (3, 1, $\frac{1}{2}$ and 3, 1, $\frac{3}{2}$) to the lowest energy level (3, 0, $\frac{1}{2}$).*

and that the direction of the spin axis could be oriented in two ways, commonly described as up and down. Going back to the example of the postal address, if all houses in the town are divided into two flats, it is necessary to add 'upper' or 'lower flat' to the address if post is to reach its correct destination.

This idea of electron spin and the fourth quantum number worked well as a means of labelling quite complicated spectra and of explaining the ordering of elements in the periodic table. There was, however, no good theoretical basis for it – the concept of spin just had to be tacked on to explain the observations.

Around the same time in the late 1920s, Paul Dirac, a brilliant young English physicist working in Cambridge, was trying to devise a new wave equation for the electron. Schrodinger's equation of 1926 was fine in that it agreed with the new ideas of quantum theory, but it did not satisfy Einstein's theory of relativity. While still only twenty-six years old, in 1928, Dirac solved the problem. To everyone's delight and surprise, the new Dirac equation required the electron to 'spin' by just the amount required to explain the observations of spectra. No longer had the fourth quantum number to be tacked on as a not very satisfactory appendage; it became a natural consequence of the mathematical description.

It is this sort of result which confirms the scientist in his belief in the existence of a deep orderliness and consistency in the natural world – an orderliness I emphasized in the early chapters of the book when pointing out that the physics of the laboratory turns out to be the same physics which is observed by astronomers to apply far removed in space and time in the most distant parts of the universe. It is this belief in order and consistency which enables the scientist to formulate principles and give them the status of scientific law. In fact, the whole of science depends on it. And not only the whole of science, but our ordinary day-to-day living depends on that fundamental consistency and order more than we readily realize.

While emphasizing the importance of order and consistency in the scientific description, it is nevertheless important to realize that scientific laws, so called, are still only *descriptions* (albeit in most cases very good descriptions) of the world as the scientist sees it. In no sense do scientific laws *make* things happen. An apple, in falling to the ground, we say, is obeying the law of gravity; but the law of gravity does not *cause* the apple to fall, it is *describing* what normally occurs.

In chapter 2, we presented a picture of God as the greatest conceivable being, the sustainer of the universe, moment by moment, keeping all its component parts in being. If that is the case, it is he who is the ultimate cause, and our scientific laws are then descriptions of God's normal activity.

With this picture in view, does not the suggestion that God might 'intervene' in the natural world seem unnecessary and superfluous? He is already there all the time. Our experience as scientists and our expectation as believers in God as creator and sustainer is that the natural world should demonstrate an extremely high degree of orderliness, consistency and stability. Where then is there room, if indeed there is, for what we commonly call the supernatural or for miracles?

First, we need to ask just what we mean by a miracle. Clearly we mean an unusual or striking event, but not just that. Many unusual events may arouse our curiosity; we may remark, 'How odd!' – but we would not describe them as miracles. In order to be classed as a miracle, the event must have significance apart from its being unusual. It must convey a message to the person or persons involved. The message may be to do with God's providence – a 'miraculous' escape or unusual provision of some kind – or it may be related to a particular physical or spiritual need or to guidance regarding a future decision or action.

In considering the occurrence of miracles, it is convenient to divide miraculous events into two categories, dependent on how they might be viewed by a scientific observer. First, there are those events which, because of their timing or unusual character, have a particular significance for an individual or for a group of people, but in which a scientist, if he were present and in possession of all the facts, would not find anything anomalous from the strictly scientific point of view or anything outside conventional scientific experience. Secondly, there are events which again have particular significance, but which it is likely that a scientist would find anomalous, with some parts of the events not fitting in with normal scientific law. Many of the miracles in the life of Jesus fit into this second category. We shall call them miracles of the second kind.

We look first at events in the first category, the unexpected meeting, for instance, or the providential escape, which, though not out of the ordinary from the strictly scientific point of view, when viewed with the 'eye of faith' (*i.e.* in the context of a personal relationship with God) take on a particular relevance or significance.

Take, for instance, the familiar journey of Paul from Jerusalem to Rome which included a storm – not unusual in the autumn in the Mediterranean – and a shipwreck on the island of Malta.[1] A scientist present would doubtless have been able to fit all the facts together into a satisfactory scientific description. Paul and Luke (the author of Acts), however, see the events as fitting into a particular divine plan.

'Is not the latter view from the perspective of faith an illusion?' the sceptical scientist might argue. After all, the events in question are following nothing but the ordinary process of scientific law. To argue this way, however, is to fall into the trap of reductionism, the falsity of which I sought to demonstrate in the last chapter. The scientist has no right to say, 'It is nothing but . . .' The description from the point of view of faith and what is seen as God's plan is complementary to the scientific one and has an equal claim to be considered. Neither description rules out the other; they are views from different perspectives.

We have seen the importance of consistency in the interpretation of the scientific view – a quality which, we have argued, stems from the characteristics of the creator and sustainer of the universe himself. The more deeply the natural world is understood, the more consistent does the scientific position appear. Consistency is also important in the view of faith. Just as the scientist achieves considerable satisfaction in seeing how events in the physical world fit together in terms of scientific law, so the Christian will find support for his faith as he is able to interpret the events around him in terms of a divine plan.

The Christian who is also a scientist is therefore looking for a double consistency. Events can and should be looked at as thoroughly as possible from the scientific standpoint. But they should also be viewed from the perspective of faith. Jesus encouraged his followers to look at all events and circumstances in this way. He chided the Pharisees, the religious leaders of the day, for their blindness; their interpretation of the sky for weather-forecasting purposes was quite good, he told them, but their interpretation of the events of the times was highly blinkered.[2]

In other words, their meteorological science was effective, but they were severely lacking in their appreciation of God's activity in the world. They were restricted by a limited and bigoted view and failed to recognize God's revelation when it came. They were blind to the message which Jesus brought and they failed to understand the core of God's plan for human beings, namely that they should be drawn into a relationship with himself – a theme we explored in chapter 6.

In some events, seeing the two consistencies for which we have argued will be relatively easy. This is not, however, always the case. The Christian believer is often faced with circumstances which seem difficult to understand. Sometimes, taken from almost any point of view, they appear wrong. Prayers do not always seem to be answered. It is on these occasions that the sceptical scientist, to whom this 'faith talk' appears in any case as wishful thinking, if not nonsense, is at his most critical. Belief in an overall consistent plan hardly seems supported by the facts. Several points need to be made about this very real problem.

First, it is important to remember that, while in the scientific description we are dealing with impersonal things, in the 'faith' description we are confronted by a personal God and concerned about a relationship in which trust is a vital element. A child will trust a parent even though he does not understand; it is the sort of trust which is the basis of faith. I must trust the controller of all, whom I am privileged to address as 'Father'. This means that the exercise of faith must never be confused with magic. <u>Prayer is not 'rubbing the lamp' and making wishes; its centre and purpose are not to get what *I* want, but to provide a means of aligning my will with God's will and of coupling my poor efforts at serving God with his substantial energy and strength.</u>

Secondly, we are bound to realize that our knowledge of God's overall plan is inevitably extremely limited. A junior infantry officer engaged in a battle may be puzzled by his particular instructions; he will possess very limited knowledge as to where his part of the action fits into the overall strategy of the battle. In a similar way, if I am a participant in God's plan, there cannot fail to be an element of mystery about where my small world and limited capability fit into it. God being so great, his plan will be grand and comprehensive far beyond the limits of my imagination.

Thirdly, it does seem that the larger the faith and the deeper the commitment, the greater the degree of understanding. The eighteenth-century poet, William Cowper, a person often given to deep depression, wrote a poem which begins:

> God moves in a mysterious way
> His wonders to perform.

He goes on to affirm:

> Deep in unfathomable mines
> Of never failing skill
> He treasures up his bright designs
> And works his sovereign will.

Our sceptical scientist, however, will still have a major objection. We live in an orderly world, he will argue, in which events follow cause and effect in a deterministic manner. How then can it also be said that these same events are moving in accordance with God's plan? It might be thought that the answer to this objection depends on the extent to which the scientific description of events is a fully deterministic one. I do not think that it does. Remember, going back to chapter 3, that God was pictured as the greatest possible designer. One way of looking at the way God runs the universe is therefore to imagine God to have taken account of all possible requirements in his initial conception. A second way of looking at it, however, is, I believe, better. In chapter 5 we pointed out that time is intimately and inextricably linked with space in the material universe. Our view of God as creator, therefore, is not only of him as creator of the spatial components of the material universe but also as the creator of time. With this view in mind, we can suppose that God fits all requirements within his moment-by-moment sustaining activity of keeping the universe going. Whatever way we try to look at it, however (and we return for another look at the problem later in the chapter), what is important is that we see God as big enough to take care, on the one hand, of the scientific orderliness and consistency, and on the other hand, at the same time, to provide for the fulfilment of his overall plan, including the place individual circumstances have within that plan. We can, therefore, be confident in our search for the double consistency for which we have argued.

We now turn to miracles of the second kind. First, consider those events in the life of Jesus which a scientific observer (if present with appropriate equipment) would probably have concluded were scientifically anomalous and not in agreement with known scientific law. Most of the miracles recorded in the gospel stories were connected either with the person of (1) Jesus (the virgin birth, the resurrection, the post-resurrection appearances and ascension), or with healing (including three (2) instances of people being raised from the dead), or with the provision of need (for instance, the feeding of the five thou- (3) sand or the turning of water into wine).

As scientists, we are bound to be sceptical about such events, which are anomalous and which do not fit into the normal scientific pattern. But also, as scientists, we are bound to pay particular attention to the seeming anomalies in case there is something of importance we have missed.

Investigation of the unexplained is in fact one of the major methods of scientific advance. The observations of perturbations in the orbits of the planets Uranus and Neptune, published by Percival Lowell in 1915, led to the discovery of the planet Pluto. The unexpected fogging of photographic plates led to Becquerel's discovery of radioactivity. In a similiar way, with regard to the unusual events during the ministry of Jesus (if indeed they occurred), it is necessary to ask whether they are conveying a particular message of which we need to take notice.

The miracles of Jesus were seen by the gospel writers, especially by John, as signs,[3] whose purpose was not only to meet a need but to *authenticate* the message Jesus came to bring. This message was that Jesus himself is a divine person – hence the emphasis given to his resurrection – and that he came to show compassion, love and forgiveness, of which the miracles of healing and provision were signs and demonstrations.

When I speak of the miracles as authentication, I do not mean that they were meant as knock-down events or 'laser'

118

miracles (as the Bishop of Durham has called them), which, by their very nature, were calculated to force people into belief. That was not the way Jesus taught and worked.[4] What I am arguing is that, when looked at in the context of his life and message, there is something of an inevitability about the miraculous events. The events are interpreted by the message, which in turn is reinforced by the events.

As we saw in chapter 6 when considering the resurrection, we can readily argue that if Jesus really is God, it is not surprising that unusual events were associated with his coming. On the other hand, how about the stress we have given, based on our scientific experience and our belief in an orderly creator, to the stability and orderliness of the natural world? If we accept New Testament miracles, is not our belief in the consistency of the universe destroyed? I do not think that it is.

Let us go back again to the Flatland illustration of chapter 5. To the Flatlanders, the appearance of a sphere from Spaceland was unusual and inconsistent with Flatland science. But looked at from the further perspective of the third dimension, it was entirely consistent. Further, a demonstration of that kind was necessary if the Flatlanders were to have any inkling of the existence of the third dimension.

In a similar way, we can look at miraculous events in the life of Jesus. From the strictly scientific point of view they are anomalous. But, just as in science the interesting anomaly can be an important lead in pointing to the way forward, so in our view of the events associated with the life of Jesus it is necessary to balance their anomalous character and their apparent scientific inconsistency against the broader consistency which appears when we look from the perspective of the spiritual dimension. Because of the demands of the spiritual view and the particular nature of the events, Christians allow that view to extend the more limited scientific view.

Consider the problem Jesus faced in getting his message across. It would have been futile for him to claim to be divine and to have a special relationship with the creator if he failed to demonstrate the claim through actions as well as saying it in words. Jesus makes precisely this point when he claimed God's prerogative to forgive, and demonstrated deity by healing a man with paralysis.[5] His actions were relevant to his message and appropriate to the expectations of the time in

119

which he lived. Standing back and looking at the teaching of Jesus about love, compassion and provision and about himself, we find a *fitness* in his actions, be they ordinary or miraculous, in giving credence to his teaching. We need such windows into the spiritual dimension if we are to take it seriously.

When this view is taken, it may seem surprising that the life of Jesus was not surrounded by many more miracles and unusual events. As we read the gospel stories, there seems to be a measured restraint. Remember Jesus was fully human as well as divine (*cf.* chapter 7). He accepted completely the limitations of a human existence and rejected out of hand temptations to throw overboard those limitations in order to serve himself.[6] Nowhere can this be seen more forcibly than when Jesus approached his death, realizing that there were means instantly available for him to avoid the shame, the pain and the cruelty of the cross.[7] However, because of his determination to accomplish the grand plan for man's salvation, he allowed himself to be cruelly put to death. As Sir Robert Boyd has rather poignantly put it, 'God holds the iron rigid and the cross upright as we impale His Son.'[8]

So far, for miracles of the second kind (those that are scientifically anomalous), we have been considering only those miracles associated with the life of Jesus. We also need to ask the question whether other miraculous events have occurred or might occur today which would now fit into a normal scientific description. The picture we have presented of God's relation to the universe clearly allows for the possibility of one-off events which, because they are one-off, do not fit the scientific pattern. The arguments are the same as those we put forward in response to the miracles during the life of Jesus. We may feel that, for many events claimed to be miraculous, however, the historical evidence is weak; in some cases too there may not seem to be a particularly strong spiritual meaning to be conveyed. For any events, therefore, that are claimed to be in this category, we shall want to look critically at the historical evidence, and to analyse it, so far as that is possible, according to criteria of fitness and consistency from both the scientific and the faith point of view. Inevitably, our limited knowledge and understanding will make it difficult, if not impossible, for us to be categorical about many cases.

Others have considered the question whether miracles happen today much more thoroughly than I can here.[9] I would just like to make two points.

First, I have constantly emphasized the orderliness of nature and the remarkable consistency of the scientific description, pointing out that these are a reflection of the character of God the creator and sustainer. We have already noted that Jesus himself resisted the temptation capriciously to interfere with this order for demonstration purposes,[10] and that, apart from miracles of healing, miracles in the sense of events outside the normal scientific order are surprisingly rare in the New Testament, or, for that matter, in the Bible as a whole. As we emphasized earlier in the chapter, the Christian, therefore, will normally be looking for evidence of God's activity in the usual circumstances of life, which, as we have seen, although entirely consistent with a scientific description, can from the perspective of 'faith' have particular significance – significance which can sometimes quite properly be described as miraculous.

My second point concerns the subject of healing, which featured strongly in the ministry of Jesus and which has always been a concern of the Christian church. Christian missions have been pioneers in bringing medical aid to all parts of the world. In recent years there has been a renewed interest in 'miraculous' healing as part of the church's ministry, as a result of which some contrast such 'miraculous' or 'faith' healing, which is seen as the work of God, with the application of medicine which is seen as man-made activity. Such a view is not only misleading, it is at variance with a belief in God as creator and sustainer of the natural order. Medical means, being derived from God's creation, are very much the works of God. Doctors, therefore, often see their role as one in which they are cooperating with God. This is even more the case as they increasingly realize the importance of treating the whole person. They cannot be concerned just with the body and its biochemistry. Complete healing involves the whole person – body, mind and relationships with others, including, I would argue, relationship with God. Such complete healing is often evident in the healing ministry of Jesus.[11] The Christian community therefore should be working together with doctors to provide a more complete healing ministry – an arrangement which brings into play all the resources God has provided for the purpose of healing. Some of the most striking modern 'miracles' are those in which both bodily and spiritual healing have occurred together.

Three further objections to the view I have tried to expound are often put forward. Before leaving the subject of the relation between the natural and the supernatural, we need to address these views briefly.

The first objection arises from a common feeling, well publicized by the French geneticist Jacques Monod,[12] that because so many events are governed by chance, there is no room for a divine plan. Monod comments regarding the origin of life, 'Our number came up in the Monte Carlo game.' It is certainly true that considerations of chance and probability play a large part in our modern scientific description, be it physics or biology.[13] It is also true that our daily lives are filled with 'chance' occurrences. We are bombarded with statistics about the probability of death from various causes, of being born with various defects, of particular sorts of crime, and so one. Insurance against improbable events is big business. But it is important to realize that 'chance' does not *make* things happen. As Donald MacKay[14] cogently argues, chance and probability, although properly part of a scientific description, are not causes of events any more than any other scientific description or law can be said to have causal properties. Further, because the scientific description of some event may involve chance and probability, it does not follow that a theological description of the same event in terms of God's activity need also involve chance. The title I have chosen for this book, taken from Einstein's exclamation, 'God does not play dice!', serves to emphasize this point. Our discussion in chapter 9 showed that the existence of a scientific description in terms of chance, however fundamental that description may appear to be, has in no way explained things away.

A second objection to the view we have put forward concerns the reality of human freedom. If all events are attributed to the activity of God, does not that make nonsense of man's freedom of choice? And if, despite all the constraints associated with any given choice, man's choices can genuinely be freely made, cannot God's plan be thwarted? In chapter 7, we considered two particular paradoxes – one in science, one in theology – which arise because of the limitations of our ability to describe views from different standpoints at the same time. Here is another paradox which arises from a similar confusion – on the interface between science and theology, perhaps the most difficult of them all. Let us look at the problem from the two standpoints, man's and God's, in turn.

Taking our point of view first, as human beings living in our four-dimensional world of space and time, we feel sure of our ground. We have no doubt about our freedom to make choices; we are making them all the time. The choices are, of course, restricted by the structure of space and time in which we exist. They are also constrained by our environment or our circumstances, and our behaviour is undoubtedly in some ways preconditioned by our upbringing and our past experience. But not entirely so; our freedom to choose is real. In particular, we are constantly faced with choices between good and evil and are only too well aware of a propensity to choose the evil rather than the good. Donald MacKay[15] has addressed the question of our freedom to choose in a formal way, and has argued that because we are self-conscious beings, the reality of our freedom of choice is demanded on logical grounds. His arguments have some appeal and were mentioned in the last chapter, but I tend to agree with John Polkinghorne[16] in being content with treating our freedom of choice as self-evident.

We turn now to attempt to look at the problem from God's standpoint, with the further dimension outside both space and time, and to ask, 'What is the Christian perspective?' Here, I want to make four points. First, if God is the creator, then it is he who has brought into being a universe in which creatures called human beings possess the power of freedom of choice – in particular, freedom to respond and freedom to love. It is God also who has chosen to limit to some degree his own freedom of action in order to allow his creatures the freedom to respond. It is the love of God which draws, not the power of God which coerces.

Secondly, we can say that God responds to man's choices, but in that response he is not constrained by space and time in the same way that we would be in making a response to action by others. We must imagine God responding to a particular human action or human prayer with freedom to act at many different times and places – in the past as well as in the present and the future. 'Before they call I will answer,'[17] God is seen as saying to his people. God's response to us is described by Jesus in terms of the relationship of a father to his children. Our Father God knows all about us (even the hairs of our head are numbered[18]), cares about us[19] and has a purpose for us.[20]

As a particular result of this view, consider a problem sometimes experienced by a person who wants to pray, say, at the end of the day about an event which he knows has occurred during the day. The outcome of that event, although unknown to the praying person, may have already happened. Does prayer after the event make any sense? Realizing that prayer is communication with a God who is not constrained by time, we can say that such a prayer is entirely appropriate. In saying that, we need also to say, of course, that were the outcome of the event already known to the praying person, it would in no way be sensible to pray for any change to the outcome; prayer cannot alter events which in our time framework have already occurred.

My third point regarding the Christian perspective of the divine standpoint is that, despite human freedom, God's purposes will be fulfilled. In attempting to describe God's relationship to activities within the world, we have frequently mentioned the idea of God having a plan. By this, we do not mean that everything is so determined by divine action that any freedom of choice we thought we had becomes illusory. It means I believe that God is somewhat like the grand master chess player, playing someone of ordinary skill. Both players are acting within the rules of the game. The grand master, however, can be said to have control of the game all the way through, even though his opponent has freedom to make his moves. The skilled master weaves in the moves of his fellow player in order to bring about a victory which is never in doubt, although the character of the end is determined by the moves of both players.

Another analogy which may be helpful is that of fault-tolerant hardware which I introduced in chapter 3, where I suggested that God, in his overall design of the universe, has built in a degree of fault tolerance. God's design and plan for the universe can take account of the fact of human freedom. Choices by human beings are real enough, but God, being the great designer that he is, ensures that, despite all, his purpose is fulfilled – in particular his purpose concerned with ultimate triumph over evil, about which we shall have more to say in the next section.

Before moving on, however, I need to point out a limitation of the analogies I have used; after all, no analogies which attempt to parallel God's activity can be very satisfactory. The analogies of the super chess player and the fault-tolerant designer suggest a God who is about the task of ensuring an end result despite the actions which human beings may take. That is, however, only part of the story. Although man's actions often are negative so far as the purposes of God are concerned, they need not be that way. Rather they can and should be positive. God invites, or rather demands, man's cooperation in *his* work. Jesus instructed his disciples to be 'the salt of the earth' and 'the light of the world'[21] and to continue the work which he had begun.[22] A further analogy we can use therefore is that we are God's fellow workers[23] or God's stewards.[24] This theme is pursued in the next chapter, where we shall address man's particular responsibility for stewardship of the earth and its resources.

Fourthly, I need to emphasize again that in all of the above, God's action is regular and orderly in a way which, we have come to recognize, expresses the divine character. As we have constantly stated, scientific observers and Christian believers alike have a strong belief in the order and consistent behaviour of the physical universe. There may remain, however, a nagging feeling that we are sweeping under the carpet some basic inconsistency. If we find ourselves still asking where in the physical system there is room for God to manoeuvre, we should also ask ourselves whether we are still too conditioned by ideas of determinism and by the reductionist approach.

It will help to recall another question which we briefly addressed in the last chapter, that of where within the scientific system can we find room for our own freedom to act – a not dissimilar problem, and one which is much nearer to us. As we saw there, we have little understanding of how our self-awareness and our freedom of action can be fitted in to the scientific scheme of things. But if, controlled by physics and chemistry as we are, we nevertheless possess the property of self-awareness and the ability to make free choices, it can surely be argued *a fortiori* that God, the greatest conceivable being, has freedom to act and respond yet all within his own constraints of order and regularity.

We have now looked at two of the particular objections to the view I have tried to expound. These have been concerned with the role of chance and the question of freedom.

Following on from these, we meet head-on the moral objection. If God is creator, sustainer and controller of all, the greatest conceivable being, how about the problem of evil and suffering? We say that God is good and all-powerful; how can he be both, yet design a universe which includes the possibility of evil and then within that universe allow apparently unrestrained evil and suffering to continue? This is the biggest issue of all which we face as human beings. Much has been said and written about it over the years. A number of points are commonly made. For instance: much evil and suffering are the result of human sin (sins of neglect as well as of positive harm). God created man to be free, the freedom to love implies of necessity the freedom to hate; to be able to choose the good implies the possibility of evil for the choice to have meaning; pain provides a necessary signal for us to take care of our bodies. Such statements may be helpful in facing the problem; they do not, however, provide an explanation.[25]

Theologians have tried to come to terms with the issue by defining different kinds of 'the will of God' – on the one hand, good events which result from his direct activity, and on the other, evil events which God allows but for which he is not directly responsible. Although it is perhaps helpful to devise terminology which prevents us from directly connecting God with evil and suffering, the distinction is bound to appear somewhat contrived. It is interesting to note that the main message of an ancient treatise on the subject, the book of Job in the Old Testament, is the recognition that God is sovereign and that he is in control even of the events associated with Job's suffering. As the book ends, and Job realizes that his view of God has been too small, he responds, 'I know that you can do all things; no plan of yours can be thwarted.'[26]

Although there seems to be no fundamental *explanation*

(even if one were provided, I suspect that, as limited, not to say fallen, creatures, we would not understand it), God has provided a *solution*. The main problem addressed by the Christian message is the fact of human sin, and the heart of the Christian message is that God in Jesus took upon himself the burden of the sin and suffering of the world when he allowed people to nail him to a cross and crucify him. Thus he commends in supreme measure the quality of God's love for human beings and introduces us to a God who is the master of transforming evil into even greater good – a transformation process which is the key to God's overall plan to which we have referred so much in this chapter. As Sir Robert Boyd has written, 'If you believe God's choice was unconquerable love, it may not help you to understand why there is pain and evil, but it may do far more to help you accept it than any number of arguments about the existence of good requiring the possibility of evil.'[27] It is not surprising perhaps that those who suffer most often turn out to be the greatest saints.

We have come back, as we are bound to do in any Christian consideration of the subject of miracles, to the events at the centre of the Christian message – the death and resurrection of Jesus. Based on the resurrection of Jesus is the Christian belief and hope in resurrection. What do Christians mean by that? An analogy I find helpful is based on the computer. Computer hardware consists of the silicon chips, the wires, the disk and tape stores, the keyboards and tape decks with which the input can be introduced, and the screens and printers which display the output. The computer software consists of the programs, which manipulate (and in sophisticated computers learn from) the input data, and provide the means to organize the output and the contents of the store. The software is of no use by itself; it needs hardware on which to act and through which to be expressed. The hardware has a limited life; in time it wears out. The software is not so perishable – it can be transferred to new hardware, although it will still bear characteristics of the hardware for which it was originally written. New, more advanced hardware can provide more scope for the software, enabling not only larger calculations to be carried out more efficiently, but perhaps providing new capabilities. Our bodies are like the hardware, providing input devices (our senses) and output devices (our limbs, speech, *etc.*) a processor and storage (our brains). Some of the software is built in from the start; it is genetically determined. Other software is continuously generated throughout our lives from interaction with our environment, with other people, and with ourselves, from our thought processes and our choices and from interaction with the hardware. Our body, in due course, like any other hardware, wears out. In resurrection, the Christian hope is for a new body which will have sufficient continuity with the old to take on board the old software and which will give us new means of expression.[28]

Some clues come from the appearances of Jesus after his resurrection.[29] He appeared, seemingly, in the same body, with the marks of the nails and the spear from crucifixion, thereby demonstrating that he was the same person. Yet his body was different; it had undergone a substantial transformation. It was not subject to the same limitations; it could appear and disappear at will. Eventually at the ascension[30] Jesus in his new body entered into heaven. A symbolic act, you say. Yes, highly symbolic, crammed full of meaning and providing a significant window for us into the spiritual dimension. How our transformation into that dimension will be achieved is beyond our current ability to imagine, but it is the Christian sure hope – described and promised by no other than Jesus himself[31] – that in due course, we shall have a part in the nearer presence of God, where pain, suffering and evil are no more, and where love reigns supreme.

---

[1] Acts 27 – 28.
[2] Matthew 16:2–4.
[3] John 2:11; 4:54; 6:14; 9:3, 16; 11:47; 12:18.
[4] See, for instance, Mark 5:43; 7:36; 8:12; Luke 16:31.
[5] Mark 2:1–12.
[6] Matthew 4:1–10.
[7] Matthew 26:53.
[8] In Carl F. Henry (ed.), *Horizons of Science* (Harper and Row, 1978), p. 16.
[9] For instance, Donald Bridge, *Signs and Wonders Today* (IVP, 1985).
[10] Matthew 4:1–10.
[11] See, for instance, Mark 2:1–12; John 5:1–15; John 9.

[12]*Chance and Necessity* (Collins, 1971).

[13]A. R. Peacocke, in *Creation and the World of Science*, chapter 3, gives examples of the appearance of order-through-fluctuations in physical and chemical systems.

[14]*Science, Chance and Providence* (Oxford University Press, 1978).

[15]*In Brains, Machines and Persons* (Collins, 1980).

[16]*One World* (SPCK, 1986), p. 94.

[17]Isaiah 65:24.

[18]Matthew 10:30.

[19]Matthew 6:8, 26, 32.

[20]Matthew 6:33.

[21]Matthew 5:13–14.

[22]Matthew 28:19; John 17:18; 20:21.

[23]2 Corinthians 6:1.

[24]Matthew 25:14–30; Luke 12:42–48.

[25]C. S. Lewis, in *The Problem of Pain* (Geoffrey Bles, 1940), provides a helpful discussion of the problem.

[26]Job 42:2.

[27]*Op. cit.*, p.16.

[28]1 Corinthians 15:35–57.

[29]Luke 24:13–49; John 20:10; 21:23.

[30]Luke 24:50–51; Acts 1:9.

[31]John 14:2–3.

Chapter Eleven

STEWARDS IN GOD'S WORLD

Where there is no vision, the people perish. (Proverbs 29:18, Authorized Version)

The human race has survived and developed through its ability to make use of the world's resources. In his agriculture, commerce and industry, man has exploited the animal, vegetable and mineral resources around him. The extent of that exploitation has grown so rapidly in recent years that the question is being raised as to how long the exploitation can continue.

Fossil fuel reserves are not inexhaustible. Some minerals are becoming much more scarce. Pollution of many kinds is a major problem. Further, demand is greater than ever before, largely because of the growth of human population – a direct result of the control of disease through medical advance.

In the past, scientists have, by and large, taken the view that they can happily pursue research and develop technology without being concerned about the danger of over-exploitation. The situation has, however, changed. It is no longer possible for the scientist to abdicate responsibility for the results of his research and the exploitation of his ingenuity. He, too, is part of the human race whose future is at stake.

The view that man is responsible is an old one. As part of the story of creation in the first chapter of the Bible, we find that

'God created man
in his own image,
in the image of God

131

                    he created him,
                    male and female
                    he created them.

'God blessed them and said to them, "Be fruitful and increase in number; fill the earth and subdue it. Rule over the fish of the sea and the birds of the air and over every living creature that moves on the ground." '[1]

A few chapters on, as part of this basis of man's activity on earth, the stability of creation is emphasized:

                    'As long as the earth endures,
                    seedtime and harvest,
                    cold and heat,
                    summer and winter,
                    day and night
                    will never cease.'[2]

Moving to the New Testament, we find Jesus emphasizing the importance of the stewardship of our abilities. Making use of and developing our capabilities is commended in the parable of the talents.[3] Particularly harsh words are reserved for the man who failed to make use of his talent and hid it in the ground.

Given a recognition that on the one hand there is a need to employ the world's resources and on the other hand to be responsible about the way they are used, we need to ask questions about the degree of exploitation that is necessary and how the responsibility should be exercised.

Because of the current, very proper, concern regarding problems of resources and of pollution, some voices are expressing strong condemnation for the attitude of Christians, especially Protestant Christians during the last century, which has encouraged (as they see it) unbridled and irresponsible exploitation of scarce resources.[4] They argue that many of our ecological problems directly result from an overemphasis by Christians on man's dominion over nature as expressed in the first chapter of Genesis, which has been interpreted as a mandate to exploit. Man therefore feels 'superior to nature,' even 'contemptuous of it, willing to use it for his slightest whim'.[5]

Other voices, driven by ecological concern, are urging

retreat from technological progress in various back-to-nature movements. A more primitive lifestyle is glamourized, areas growing wild are preferable to cultivated ones, pesticides and artificial fertilizers are to be eschewed, big industry should be fragmented into much smaller units. Some even argue that just as Prometheus suffered for stealing fire from the gods, the curses of technological progress are nature's revenge for man's improper prying into nature's secrets. Not dissimilar attitudes, antagonistic to science and technology, are held by many Christians who, worried about overemphasis on material gain, stress the dominant importance of the 'spiritual' and look for spiritual solutions to material problems – for spiritual healing independent of any medical treatment or for guiding messages from heaven divorced from more down-to-earth means of arriving at decisions.

Although the contributions of science and technology have been essential (and they are still very much required) in helping to provide for the world's basic needs, these reactions against technological progress are understandable. We can all think of ecological disasters brought about by irresponsible exploitation, often defended on grounds of man's God-given right to use the world as he pleases. Widely publicized studies of global ecology[6] have emphasized these problems. These studies, however, have not gone without criticism. A critique of Global 2000[7] has pointed out its one-sided view, that nature is more resilient than appears at first sight, and that man's ability to adapt tends to be underestimated on the simple assumptions of the global studies. Professor Lovelock[8] goes so far as to see man together with his environment as forming a single complex system which itself defines and maintains conditions necessary for its own survival.

What is clearly needed in this debate is integrity and balance, and the establishment of principles on the basis of which sensible planning can be pursued. How can the Christian help? What principles can be proposed on guidance for the future?

The bases of the Christian view are, first, that it is God's world in which we live – he is both creator and sustainer of it – and, secondly, that man has been made in God's image, with some God-like qualities, and placed in the world to be its steward. Man, therefore, should not be afraid to grasp as

133

Fig. 11.1 *The Olympus satellite – a large communications satellite due to be flown by the European Space Agency, to provide direct broadcast television and business services over Europe. One channel of Olympus' communication package can*

transmit information at a rate equivalent to more than 'one Encyclopaedia Britannica' *per second* (courtesy of British Aerospace).

grasp as

thoroughly as possible the resources and the capabilities he has been given, using them first to express worship for his creator (a subject we shall pursue further in the next chapter), and secondly to care for the world and the human beings within it in ways which are consistent with the declared wishes and purpose of the one for whom he is acting as steward. Such was very much the attitude of the early scientists who fostered the rise of experimental science in the seventeenth century and who pursued their science and developed their technology for the glory of God and for the benefit of mankind.[9]

An essential emphasis of the Christian view is that man is steward under God. No-one therefore has the right or mandate to exploit nature for his own glorification or agrandisement or solely for his individual benefit. The requirements for man to love God and to love his neighbour as himself[10] need to be expressed in the way man exercises his stewardship over the world and its resources. I would like to suggest three criteria which should govern man's stewardship.

First, there is the pursuit of fulfilment. Jesus told his disciples that his purpose was that they might have life to the full.[11] How can man, made in God's image, develop his God-like qualities?

I suggest he can do so by his enjoyment of the natural world, by being creative within it, and by facing its challenges.

Many expressions of pleasure in the created world occur in religious and secular literature. One from the Old Testament is carved over the door of the Cavendish Laboratory in Cambridge: 'The works of the Lord are great, sought out of all them that have pleasure therein.'[12] It is clear also from his use of many illustrations and parables from nature that Jesus found a great deal of enjoyment in the natural world. Even the rather ascetic apostle Paul tells the readers of one of his letters that God has given us richly all things to enjoy.[13] Scientists of past centuries, in their books and papers, often expressed their enjoyment of their work. Scientific literature today is more formal and turgid; nevertheless, enjoyment remains an important motivation in scientific and technical work – enjoyment which the scientist has a responsibility to share with others who, though not actually engaged in scientific pursuits, can appreciate its challenge and its results.

Fig. 11.2 *The man-machine mix. A Meteorological Office forecaster has access by computer terminal and display to data (including satellite and radar images) and computer-generated forecasts. To these he adds other information and his own knowledge and experience to provide his weather forecast.*

To be creative like God is a basic human ambition. The scientist and technologist have their fair share of creativity. Those who develop and use their ideas in the workaday world, however, have little chance to be creative. They see the curse of automation consigning them to a life of monotony or of unemployment. One of the challenges of the day is to see automation as a tool, not a master, and to develop jobs and working methods in which man continues to employ his creative abilities to the full while making maximum use of the powerful new tools at his disposal.

Information technology[14] is a powerful new area of scientific and technical activity, covering the worlds of communications, computing, artificial intelligence, automation and robotics. Human activities of many kinds, in the workplace, the home and the leisure centre are gradually being transformed by the new possibilities information technology is opening up. Much attention is being given to the man–machine interface – the development of ways in which the man and the machine (be it computer or robot) can help each other. The potential is there in these new technologies for man's creative abilities to be extended, and for more people in our industries to feel the satisfaction of creative employment.

Finally, there is the fulfilment that comes through facing the challenges of science and technology, not just those of prising out nature's secrets, but also the challenges of guiding the applications of the knowledge gained to the benefit of mankind.

A second criterion governing man's stewardship is that its aim should be an increase in human freedom – freedom from disease, from oppression and from drudgery. Care for the poor and needy receives high priority in the Bible. Jesus stressed the importance of such care in his teaching and amply demonstrated his determination to overcome sickness and disease. He also expressed concern that people should be sufficiently free from care and worry to have time for other things, especially to have time for activity in the spiritual dimension. He emphasized that 'man does not live on bread alone'.[15]

It is relatively easy to talk of a greater realization of human freedom as a rather utopian ideal. It is much harder to turn such principles into practical reality, involving, as they do, not just technological considerations, but also political and economic ones.[16]

It is not appropriate here to go into details, except perhaps to say that it is clear that progress in this direction is bound to imply a more equable sharing of world resources, greater assistance by developed nations with appropriate technology for the developing nations, and the development of a greater ability to forecast or prevent natural disasters and to alleviate their consequences.

The third criterion I wish to emphasize as governing man's stewardship is that of foresight for the future. Almost universally man is concerned for the well-being of the immediate future generation; we all want to see our children properly established. But our concern becomes less acute as we look to the generations beyond. 'We have managed our situation,' we say, 'they can learn in their turn to manage theirs.' But is that adequate stewardship? We need to look at it not only in the context of future human need but also in the context of the long-term management of the planet and its resources, biological, chemical and physical. It would be a poor farmer who impoverished his land for quick, easy profit; we would be inadequate stewards if we failed to consider the long term.

Long-term considerations have become more acute as man's activities have grown to the level at which large changes, for good or ill, can rapidly be brought about. Our responsibility as stewards becomes that much greater. Further, because so much information is available, as is instant communication with almost any part of the world, a global picture of most problems can be perceived. The challenge, therefore, is presented of finding global solutions. Questions about world energy policy need to be asked. Ought we to be planning to preserve more of the world's oil for future use? It is, after all, a valuable chemical resource as well as an energy one.[17]

Questions regarding population growth also come up. Modern medicine has done well in ensuring lower infant mortality and a longer life span. Can the resulting population explosion be sustained? If not, how can it be controlled? Good stewardship demands that these global problems be faced and if necessary global solutions proposed.[18] We have been considering man's responsibility as a steward for the world in which he lives. In earlier chapters, we emphasized the

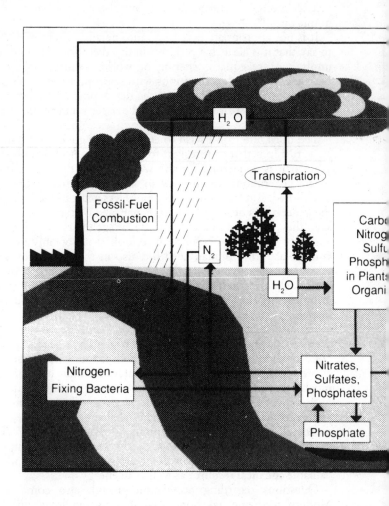

Fig. 11.3 *Schematic representation of some of the complex biogeochemical cycles, i.e. cycles in which chemical and biological processes within the atmosphere, oceans or land surface interact with each other. Of particular interest and concern is the impact of human activities on these processes. For instance,*

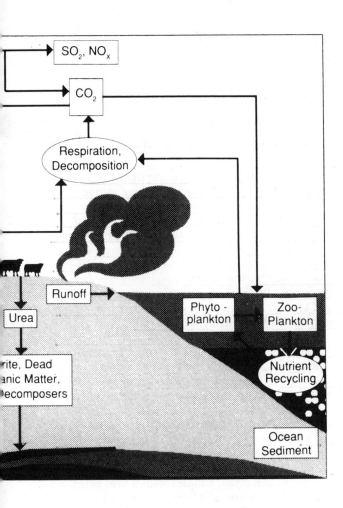

the burning of fossil fuels, oil, coal, wood, etc., leads to an increase in atmospheric carbon dioxide which in turn, on a timescale of a century or more, may lead to significant climate change.

141

control which God exercises over the universe as its sustainer; God's continual activity keeps the universe in being. The question might well then be asked, 'If God is in control, where does man's stewardship come in?' Is the delegation of stewardship to man a genuine delegation?

As we saw in chapter 10, the statements that on the one hand man has freedom to act and on the other God is in control should not be seen as incompatible; they are statements at two different levels in the hierarchy of description (*cf.* chapter 9). Remember that God is outside our space and time.

But, as we also saw in chapter 10, it is important to realize that man's freedom and God's control are not independent of each other. Man is not meant to act independently of God but rather cooperatively – as indeed the word 'stewardship' implies. We are to seek out and do the will and intention of the one for whom the stewardship is being carried out. In the parlance of the dimensional argument of earlier chapters, man is not just to act within the four-dimensional universe of space and time; he is also to link up with God in the spiritual dimension and set his perspectives and activities in a world of full dimensionality – spiritual as well as material.

In practice this means the exercise of stewardship in as thorough a manner as possible (under the sort of criteria we have outlined), looking to God in the spiritual dimension for moral and spiritual strength to carry it out. It also implies a realization that surrounding and above our activity is God's activity. For any given situation, there are bound to be limitations to our knowledge and our ability to control; but what we are invited to do is to go into the situation in partnership with God, knowing that he can take care of those things which we cannot. A clear understanding of the responsibilities and the abilities we have been given, coupled with a firm belief in God's presence and providence, is the mixture that makes Christian stewardship a challenging and exciting activity.

In the Old Testament in what is known as the wisdom literature, this prescription goes by the name of 'wisdom'. 'In his heart a man plans his course, but the Lord determines his steps.'[19] This is a typical statement expressing the joint activity of man and God. The idea of wisdom there, however, is more than just a combination of common sense and dependence on

God. It is sometimes given a capital W and is personified as if part of God – ideas which are linked with the New Testament description of Jesus being the Word of God.[20] It is through the person of Jesus that our link with the spiritual dimension is forged, and through him as God's Wisdom that our activity as stewards can be guided and controlled. A prayer of Reinhold Neibuhr, an American theologian who lived earlier this century, is appropriate as we face the demands and perhaps seemingly impossible challenge of global stewardship:

'O God, give us Serenity to accept what cannot be changed;
Courage to change what should be changed;
And Wisdom to distinguish the one from the other.'

---

[1]Genesis 1:27–28.

[2]Genesis 8:22.

[3]Matthew 25:14–30.

[4]See, for instance, L. White, 'The Historical Roots of our Ecological Crisis', *Science* 155 (1967), pp. 1203–1207.

[5]*Ibid.*

[6]For instance, D. M. Meadows *et al.*, *The Limits to Growth* (Universe Books, 1972); *The Global 2000 Report to the President of the United States* (Pergamon Press, 1980).

[7]J. L. Simon and H. Kahn, *The Resourceful Earth* (Basil Blackwell, 1984).

[8]J. Lovelock, *Gaia: A New Look at Life on Earth* (Oxford University Press, 1979).

[9]A view expounded in detail in R. Hooykaas, *Religion and the Rise of Modern Science* (Scottish University Press, 1973).

[10]Matthew 22:37–39.

[11]John 10:10.

[12]Psalm 111:2 (Authorized Version).

[13]1 Timothy 6:17.

[14]For a discussion of the detail and challenge of information technology, see T. Forester, *The Information Technology Revolution* (Basil Blackwell, 1985).

[15]Matthew 4:4; see also Matthew 6:33; Luke 10:38–42.

[16]See *North–South: A Programme for Survival*, the Report of the Independent Commission on International Development Issues under the chairmanship of Willy Brandt (Pan, 1980).

[17]See, for example, I. Blair in John Stott (ed.), *The Year 2000* AD (Marshall, Morgan and Scott, 1983), chapter 4.

[18] A useful discussion of these issues is in *Shaping Tomorrow* (Home Division of the Methodist Church, 1981).
[19] Proverbs 16:9.
[20] For a fuller development see James Houston, *I Believe in the Creator* (Hodder and Stoughton, 1979), chapter 7.

Chapter Twelve

FROM WONDER TO WORSHIP

I came to know God experimentally. (George Fox)

It is commonplace to comment on the enormous strides which have been made this century in science and technology. In our industry, in our homes, we are surrounded by the fruits of human achievement. Work and leisure alike are dominated by technological devices which press themselves upon us demanding our attention. As Ralph Emerson said many years ago, 'Things are in the saddle and riding mankind.'

No-one has described this takeover by 'things' more eloquently than Theodore Roszak,[1] who talks of science with its reductionist approach 'undoing the mysteries', destroying imagination and creating a desert of the human spirit. What he calls the modern technocracy is, he claims, succeeding by emphasis on the 'systems approach' in its 'coca-colization' of the world. Imagination, he argues, has been replaced by cold calculation in the over-objective emphasis of science.

As we have already mentioned in the last chapter, a great deal of disillusionment with science and technology has resulted from this popular perception of the scientific enterprise. The instant solutions to human difficulties, the technological fixes, so often fail to tackle the real problems or they merely succeed in replacing one set of problems by another more intractable set. Hence the emergence of a counter-culture. As reactions against the objectivity of science and the manipulative possibilities of technology, we see a vastly increased interest in magic, superstition (astrology has never been so popular) and the identity-dissolving thought of eastern

145

religions. Roszak for his part believes the solution lies in the development of a mystical approach, its content undefined and unimportant but dominated by imagination.

So far in this chapter, science and technology have been lumped together – deliberately so because in the popular mind they are perceived together. Science provides the basic knowledge through which technological developments can take place. Public support and public finance for science are more and more dependent on its applications, on its potential benefit to the economy or to equipment for national defence. Science for its own sake is looked upon with suspicion; politicians want to see scientists more subservient to their particular goals. One result of science being viewed through the technology which it produces is the disillusionment and dissatisfaction I have mentioned.

It is this confusion between science and technology which has helped to provoke such a strong reaction from people like Roszak. He believes that when the scientist speaks of beauty and wonder, he is referring to the beauty of the discovery of the appropriate formula, the solved problem, the wonder of the world 'packaged up, mastered, brought under control'. This he contrasts with the beauty of the 'magical presence' experienced by the artist and the poet who is 'awed not informed.'[2]

I believe this view of science is, however, a highly distorted one. Perhaps because of the cold, calculating way in which science is put over in our technologically advanced world, Roszak fails to realize that awe, wonder and mystery remain strong components of the scientific enterprise with its unsolved problems as well as its packaged formulae and solutions, with its concerns for qualities and values as well as for progress.

Go back to the attitudes of those scientists of 300 years ago in the early days of the advance of experimental science and we find that their inspiration came from a spirit of curiosity and from a desire to explore the works of God. A prime aim of their scientific activity was to bring glory to God. Science for them was a sacramental activity.[3]

Later scientists right up to the present day have been motivated by similar attitudes. They may not all have expressed it in the same way or seen God in it, but they have felt awed by the vastness and complexity of the universe, and they have found pleasure in the fascination, the orderliness and the stability of the natural world.

A belief in the existence of objective truth to be searched out has been and still is a powerful influence on the community of scientists. The attitude of humility before the facts, commended by Thomas Huxley, rightly pervades their work and controls their thoughts. Scientific enquiry is not devoid of mystery, awe and wonder. Let me quote from Sir Fred Hoyle, the eminent astronomer, from the conclusion of an address he gave in Cambridge:

'My object is to give you a very brief outline of the way that scientific enquiry can be brought into relation with religion. It will be sufficient if a breakthrough, however small, has been made ... I would remind you of Laplace's super-mathematician, who, you may remember, was capable of working out in full detail all the consequences of the laws of science. Now imagine an intellect to whom this would be a comparatively trivial exercise, an intellect who is interested in the consequences not of just one specified system of laws, but in examining all systems of law with a view to devising the one most pregnant with possibilities, an intellect who is able to relate the design of the laws of nuclear physics to the conditions that operate inside the stars, and who can relate the origin of the stars and planets to the intricate chemical details of the origin of life. Imagine that this is done not by issuing an arbitrary fiat – "Let there be light" – but by a complete mastery of all details of the situation. Imagine the intellectual interest and magnitude of such a problem. Then I think you will come as near as we can come, in our present inadequate state of knowledge, towards understanding the meaning and purpose of the universe.'[4]

Hoyle's scientific instinct is clearly associated with a deep feeling of wonder. Charles Coulson, who during his productive scientific life held in turn chairs of physics, mathematics and chemistry, went much further. He saw his scientific work as sacramental.[5] Like John Ray of the seventeenth century, Coulson found science a 'fit subject for a Sabbath day', and, like the astronomer Johann Kepler, in his scientific work he received great stimulus from the thought that he was 'thinking God's thoughts after him'. 'The reality of God', he writes, 'affects every issue, since whatever we see, wherever we look, whether we recognise it as true or not, we cannot touch or handle the things of earth and not, in that very moment, be confronted with the sacraments of heaven.'[6]

Coulson goes further still; he considers the qualities which enable the scientific enterprise to flourish as of Christian origin. 'When we think of the flowering of the human intellect in the humility, patience, imagination, one-ness and splendour of modern science, then we should agree not only that science is a moral enterprise, but that it holds within itself the very stuff of religious experience. And so, since the Order of Physical Nature is one aspect of God showing Himself to His children, what they see and do when they study it is most intimately bound up with what He is and what they are.'[7]

An important element in the attitude of scientists like Coulson is the idea of transcendence. By that I mean the belief that in science we are dealing with something that is 'given'. The scientific world we are describing is given by Another. This idea of transcendence is one which man has tended to lose. Roszak writes, 'Protestants still acknowledge a minimal God' – a 'stuffed God in the corner'[8] – but their God has little relation to their technological world. With the enormously powerful tools man has developed, modern man attempts to see himself as the controller of his destiny.

Theologians talk about the fall of man, referring to the biblical story in which Adam and Eve in the garden of Eden took fruit from the tree of knowledge of good and evil which they had been specifically forbidden to do.[9] By so doing man was attempting to take for himself what can only be given. Man, fallen man, still continually acts in the same way. Nowhere is this more clear than in his unrelenting pursuit of power and control through technology. Rather than acknowledging his dependence on the universe and his subservience to its creator, man attempts to set himself up as the transcendent being from whom all else flows. In terms of our dimensional analogy, man sees himself, with increased dimensionality, as the controller and operator of a spiritual dimension.

As we have seen, this attitude, characteristic of fallen humanity, colours the approach to science and technology by

Fig. 12.1 *Michael Faraday's laboratory at the Royal Institution about 1835. Faraday was one of the greatest experimentalists of all time. Our modern electrical power industry directly results from his work.*

scientists and by non-scientists, by Christians and non-Christians alike. How can we achieve a more balanced view?

First, it is important that technology be seen as a fruit of science rather than apart from it. We sometimes talk of 'pure' science, meaning fundamental science for which no immediate application can be seen. Experience shows, however, that it is rare for it to be long before fundamental research illuminates some applied problem or before it is exploited for a technological end. The two cannot and should not be divorced.

Secondly, there needs to be an increased public awareness of science and the scientific method. We are familiar with television programmes such as *Tomorrow's World* which excite the public about future applications and the possibilities for new gadgetry. People at large, however, know little about the method of science, the search for truth, the failures as well as the successes, the unsolved problem as well as the elegant solution, the value judgments as well as the drive for progress. It is important that for scientists and non-scientists alike, in our schools and colleges, the basic stuff of scientific thought and the ingredients which have led to important scientific discovery should be exposed.

Thirdly, emphasis needs to be put on the quality of transcendence inherent in the scientific enterprise, that science is dealing with things that are given. Attitudes of awe, wonder, and humility before the facts are essential if man is to be in harmony with both his environment and his creator.

Awe, wonder and humility go so far; they concentrate on the magnificence of the facts. Worship goes further, in that it addresses the magnificence of the person behind the facts.[10] I have spoken several times about the two books of God's revelation – the book of nature and the book of the Bible, the latter particularly bringing the revelation of God in Jesus. To me, the most worship-evoking experience comes from putting these two revelations alongside each other. As I turn over the pages of an atlas of astronomy which captures in colour the structure of planet or galaxy as viewed through telescope or from space probe, or as I look at the ever-changing hues of planet Earth seen from an orbiting space laboratory or from an automatic camera in geostationary orbit, I find that not only am I filled with excitement of an intellectual kind on realizing the scale and intricacy of the processes exposed by those pictures, but my imagination and other emotions are aroused with wonder and humility before the scientific facts.

Couple all that with the realization that the creator God of the universe is also the redeemer God, who has conceived a plan not only for the construction and maintenance of the physical universe, but also for the transformation of individual creatures of *homo sapiens* into his own likeness. And the enaction of that plan is not through coercion by the arbitrary exercise of power but through the inescapable attraction of God himself. Jesus of Nazareth, who lived as a man among us, allowed himself to be cruelly tortured and killed. Because of that, and because through his resurrection he demonstrated victory over evil and death, he is the focus of the power by which the transformation occurs. Instead of being selfish spoilers and exploiters, human beings, redeemed and transformed by Jesus, exhibit qualities of 'faith, hope and love'.[11] In other words, in addition to being creatures of earth, they become aware of and strongly connected to God in the spiritual dimension.

A frequently quoted statement of Jesus is, 'Seek first [God's] kingdom and his righteousness, and all these things will be given to you as well.'[12] The things Jesus was referring to were the material things of clothing, food and drink. He was not, I think, saying to his disciples, as one might say to a child, 'If you behave yourself by doing so-and-so, I will give you some sweets or other reward at the end.' Rather he was expressing something much more fundamental, a universal lesson that concentration on material things never leads to satisfaction, that it is only in the pursuit of the higher things of 'God's kingdom' that real appreciation of the material can result. As C. S. Lewis has put it, 'You cannot get second things by putting them first; you can get second things only by putting first things first.'[13]

Think for a moment about the appreciation of depth which is contained in a scene viewed with both eyes rather than with one eye, or the depth which stands out in pairs of pictures viewed through a stereoscope. Objects appear solid; estimates of distance can be made. Illusions of depth can be created by various means in pictures on paper or on a television screen, but they are effective only as they can be related to the experience we have in real life of viewing scenes with both

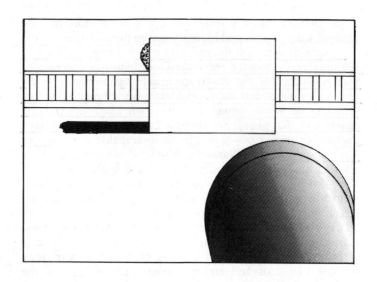

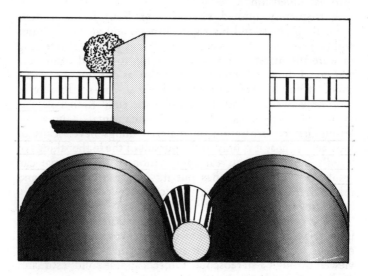

Fig. 12.2 *'Putting the two revelations of God together is like having binocular vision.'*

eyes. Putting the two revelations of God together is like having binocular vision. A new depth and reality are created; the new spiritual dimension stands out more clearly.

In this and the preceding chapters, I have indulged in a considerable amount of God-talk. I have made constant reference to a spiritual dimension. I hope I have demonstrated its relevance as a model to aid our thought and as a help in the exercise of faith. But, you may say, I have not proved anything, and how, you may ask, can faith begin? It starts in very small ways; that is the answer Jesus gave. It is like a very tiny seed – a grain of mustard seed – from which a large tree will grow.[14]

The growth is not unlike the progress of a scientific experiment, which begins with a scientific idea. An idea is suggested, perhaps from elsewhere in science; a theory is developed which is attractive because it fits in with other pieces of theory and of scientific knowledge. Suppose (the scientist argues) the theory is true; what consequences does it have? How can I test these consequences? So experiments are initiated or observations made to attempt to test the truth or falsity of the theory. So it is with the progress of faith. It starts perhaps somewhere in the intellectual framework I have been presenting, with the idea of God, or perhaps more often with the recognition of human need. The possibility that there is a God who cares about the world and who is looking for a relationship with human beings takes on immense significance. As in a scientific experiment there needs to be commitment, not just the intellectual commitment of the dedicated scientist, but rather the kind of commitment involved in a human relationship. After all, the God we have presented is a person to whom we can relate. In the Christian experiment (or should I rather say experience?), commitment may be tentative and hesitant at first. It becomes more solid as the relationship grows, eventually including all aspects of personality.

As with human relationships, the genuineness of the relationship is tested in a variety of ways, many of them of a highly personal kind. It is in the exercise of worship, prayer and hearing the word of God in the Scriptures, both on an individual basis and corporately with the Christian community, that the relationship is developed and the testing occurs. Is prayer meaningful? Are my prayers 'answered'? If the answer I want is not forthcoming, do I accept the will of God, whatever that turns out to be? Is the relationship effective in achieving victory for love, forgiveness and compassion over against failure, selfishness and evil? Can others see evidence of my being 'shaped to the likeness of God's Son?'[15]

Of course, there is nothing either trivial or easy about the process of getting to know God. I have tried to draw parallels between it and the scientific enterprise. I have suggested models with connections in science – such as that of a spiritual dimension – to help with this. As we saw in chapter 5, however, there is a danger here that, having achieved (so we think) some description of God and where he is, we have thereby put God in a box and cut him down to size. Have we just succeeded in constructing a 'graven image' of him which can be located and circumscribed?

Earlier in the chapter we saw the distorted and limited view of science which results from seeing it as mirrored by technology; the device and the gadget eclipse the enterprise and creative thought from which they are a by-product. The same danger exists with any picture images or model of our own creating which we may employ in our thought about God. They are to be used to stretch our mind and our thought; if they in any sense confine or limit God, they are doing the opposite of what is intended. It is essential to realize that 'now we see but a poor reflection as in a mirror; then we shall see face to face. Now I know in part; then I shall know fully, even as I am fully known.'[16] About the extent of that knowledge to which we can look forward we can hardly speculate, except in humility and dependence in order to pursue, through the exercise of 'faith, hope and love',[17] that partial knowledge of God which is available.

One final word, however, about both the partial view and the fuller view which is of great importance. Both views combine the material and the spiritual. Our current culture over-emphasizes the material. Many Christians and other religious groups, reacting against this, reject the material as transitory or even evil in favour of the lasting nature of the spiritual. But the Bible is a very earthy book: spiritual truth and experience are delivered through material means. The teaching of Jesus is a remarkable combination of the material and the spiritual. A central statement of the Christian creed is the belief that the body of Jesus was resurrected – a body, albeit a transformed body, in which he has gone into heaven. Of the nature of that transformation from this world into the greater world beyond, or of how the whole creation, which is groaning as in the pains of childbirth,[18] can take part in it, we have little knowledge – although the analogy of the spiritual dimension linking with the four dimensions of space and time has, I hope, helped a little in stretching our thought and imagination. In the meantime, I believe we can pursue both the quest of science with its elegance and rationality, and the quest of faith with its goal of knowing God himself, realizing that our discovery is never complete. What we discover most of all, both in science and faith, is how much more there is to find out.

---

[1]In *The Making of a Counter-Culture* (Faber and Faber, 1968) and *Where the Wastelands End* (Doubleday, 1972).

[2]*The Making of a Counter-Culture*, p. 252.

[3]For an exposition of the scientific and religious attitudes of early scientists, see Colin A. Russell, *Cross-currents: Interactions between Science and Faith* (IVP, 1985).

[4]M. Stockwood (ed.), *Religion and the Scientists* (SCM Press, 1959), pp. 65–66.

[5]Charles A. Coulson, *Science and Christian Belief* (Oxford University Press, 1955).

[6]*Ibid.*, p. 102.

[7]*Ibid.*, p. 63.

[8]*Where the Wastelands End*.

[9]Genesis 3.

[10]C. S. Lewis, in *Reflections on the Psalms* (Fontana, 1961), chapter 9, expounds the idea of worship.

[11]1 Corinthians 13:13.

[12]Matthew 6:33.

[13]C. S. Lewis, *First and Second Things* (Collins, 1985).

[14]Matthew 17:20.

[15]Romans 8:29 (New English Bible).

[16]1 Corinthians 13:12.

[17]1 Corinthians 13:13.

[18]Romans 8:22.

Index

Abbott, Edwin, 64–65
Adam and Eve, 148
analysis, tool in science, 99–100
Anselm of Canterbury, 45
anthropic universe, 46
ascension, 118, 128
Aspect, A., 101
Athanasius, 88
automation, 138

Bacon, Sir Francis, 76
Becquerel, radioactivity, 118
Bell's theorem, 103
Bible, 76–77, 95, 121, 131, 138,
    150, 154
Big Bang, 15ff., 22, 32, 35, 41
binocular vision, 153
biogeochemical cycles, 140
Bohr, Niels, 111
Boyd, Sir Robert, 120, 127
brain, compared with computer,
    107

carbon, formation of, 27, 35
Chalcedon, Council of, 88
chance, 11, 122
church, 81, 121
climate change, 139
commitment, 95, 117, 153
communications satellite, 134
computer, 107, 127

hardware, 43–44, 107
    software, 107, 127
consistency, 113–115, 118ff., 125
    double, 115, 118
Coulson, Charles, 147
counter-culture, 145
Cowper, William, 117
Crab Nebula, 22, 29, 47
creation
    special, 12
    story of, 131
crucifixion of Jesus, 81, 120,
    127–128

Darwin, Charles, 76
description, scientific, 113, 122
determinism, 125
deuterium D, 58
dimension(s)
    of space, 47, 51
    spiritual, 63ff., 66, 70ff., 76,
        79, 81, 95, 119–120, 128,
        138, 142–143, 148, 151, 153
    three/third, 50, 119
    time, 51, 56–57, 62, 64
    two, 50, 64, 79
Dirac, Paul, 88, 113
Doppler shift, 32

ecology, 133
Einstein, Albert, 11, 13, 92, 94,

157

DATE DUE

Demco, Inc. 38-293